Biodiversity, Land-use Change, and Human Well-being

BIODIVERSITY, LAND-USE CHANGE, AND HUMAN WELL-BEING
A STUDY OF AQUACULTURE IN THE INDIAN SUNDARBANS

Kanchan Chopra
Preeti Kapuria
Pushpam Kumar

OXFORD
UNIVERSITY PRESS

YMCA Library Building, Jai Singh Road, New Delhi 110 001

Oxford University Press is a department of the University of Oxford.
It furthers the University's objective of excellence in research,
scholarship, and education by publishing worldwide in

Oxford   New York
Auckland   Cape Town   Dar es Salaam   Hong Kong   Karachi
Kuala Lumpur   Madrid   Melbourne   Mexico City   Nairobi
New Delhi   Shanghai   Taipei   Toronto

With offices in
Argentina   Austria   Brazil   Chile   Czech Republic   France   Greece
Guatemala   Hungary   Italy   Japan   Poland   Portugal   Singapore
South Korea   Switzerland   Thailand   Turkey   Ukraine   Vietnam

Oxford is a registered trademark of Oxford University Press
in the UK and in certain other countries

Published in India by Oxford University Press, New Delhi

© Institute of Economic Growth 2009

The moral rights of the author have been asserted
Database right Oxford University Press (maker)

First published 2009

All rights reserved. No part of this publication may be reproduced, or
transmitted in any form or by any means, electronic or mechanical,
including photocopying, recording or by any information storage
and retrieval system, without permission in writing from
Oxford University Press. Enquiries concerning reproduction outside
the scope of the above should be sent to the Rights Department,
Oxford University Press, at the address above

You must not circulate this book in any other binding or cover
and you must impose this same condition on any acquirer

ISBN-13: 978-019-806021-5
ISBN-10: 019-806021-1

Typeset in Adobe Garamond 11/13 by Jojy Philip
Printed in India at De Unique, New Delhi 110 018
Published by Oxford University Press
YMCA Library Building, Jai Singh Road, New Delhi 110 001

Contents

Foreword	ix
Preface	xi
List of Tables, Figures, and Maps	xiv
List of Abbreviations	xvii

1 Ecosystems, Ecosystem Services, and Aquaculture: Drivers of Change … 1

- Setting the Stage: Natural Capital, Resource Use by the Economic System, and Ecosystem Services … 1
- Setting the Stage: Approaches to Human Well-being … 4
- Major Drivers of Growth in Aquaculture in the World Economy … 6
- Environmental Impacts of Aquaculture … 7
- Ecological Footprint of Aquaculture … 9
- Trade in Shrimp Aquaculture: Global and National Scenario … 12
- Trade Regulations Concerning Aquaculture Industry … 14
- The Sundarbans: A Biodiverse, Fragile, and Changing Ecosystem … 16
- Structure of the Book: A Preview … 18

2 Trade Liberalization and Shrimp Exports … 21

- Increasing Openness of the Indian Economy in the Nineties: The Macro Picture … 21
  - Exchange Rate Movements from 1991 to 2003 … 21
  - Factors Impacting Marine Products Exports from India and West Bengal … 23

India and the World Shrimp Economy		27
Marine Products and Shrimp Exports from India		30
Frozen Shrimp Exports from West Bengal		36

3 **Determinants of Shrimp Export from India and West Bengal: Analysis and Some Econometric Explorations** — 44

Introduction	44
Magnitude and Destinations of Exports of Shrimp from West Bengal	45
Model Specification and Methodology	49
The Model	49
The Database	52
Description and Construction of Variables	52
Model Results and Interpretation	54
Discussion and Concluding Remarks	57

4 **Shrimp Exports and Aquaculture: The Region and the Stakeholders** — 85

Introduction	85
The Sundarbans Region—Background and Overview	86
Socio-economic Profile of the Sundarbans	90
Stakeholders in Shrimp Production	93
Processing and Export Units	94
Shrimp Farmers	95
Farm owner-cum-worker	96
Farm Workers	98
Prawn Seed Collectors	98
Aratdars	99
Input Suppliers (feed, seed, antibiotics, tractors, pumps, aerators, and generators)	99
The Political System	100
Private Investors	101
Post-harvest Production Links	101
Local Transporters	102

Local Markets (Domestic Consumption)		102
National Government		103
Concluding Remarks		103

5 Biodiversity Loss off the Sundarbans Coast: Magnitude, Cost, and Impact — 108

Introduction	108
Indices of Ecological Crop Loss	110
Internalizing the Cost of Biodiversity Loss using Translog Cost Function	113
The Issue and the Methodology	113
The Database	116
Results and Analysis	120
Model with Biodiversity Cost Internalized in the Input Cost for Seed	120
Economies of Scale	124
Concluding Remarks and Policy Implications	125

6 Land-use Change in the Sundarbans — 132

Land-use Change	132
Evidence from other Studies	134
Methodological Framework for Land-use Change in the Indian Sundarbans	140
Objective of the Analysis	140
Conceptual Framework	140
Methodology	141
Model Specification	142
Data and Variables	144
Description and Construction of Variables	152
Empirical Estimation of the Land-use Change	155
Conversion of Paddy Land to Aquaculture (Log-linear Specification)	155
Conversion of Mangrove Land to Aquaculture (Log-linear Specification)	157
Concluding Remarks	159

7 Human Well-being: Who Gains and Who Loses?	170
Introduction	170
Poverty, Well-being, and Ecosystem Services: The Literature and Some Conceptual Approaches	170
Shrimp production and Export in the Sundarbans, Ecosystem Services, and Human Well-being	176
Income Generation from Shrimp Exports in the Sundarbans	178
Human Well-being Indices for Different Stakeholders	181
Human Well-being Indices	183
Shrimp Farmers and Agricultural Farmers in Canning and Minakhan	184
Shrimp Farmers and Mixed Income Households in Canning and Minakhan	186
PL Collectors and Fishermen in Gosaba	188
PL Collectors and Salary-wage Earners in Gosaba	190
Concluding Remarks: Income Generation, Well-being Levels, and Resource Use in Shrimp Culture	192
8 International Drivers and Local Impacts: Responses and Interventions	204
Recapitulation	204
Policy Responses: Reviewing the Concept	207
Evolution and Listing of Individual Agent Focused and Integrated Responses: Present and Future	209
Concluding Remarks: Moving Towards Sustainable Development in the Sundarbans	215
References	259
Index	267

Foreword

This volume is being published as part of the Institute of Economic Growth (IEG) series on 'Studies in Economic and Social Development' (Number 70 of the series earlier called the 'Studies in Economic Development and Planning'), being the outcome of research undertaken by its authors, at the IEG. Its subject matter falls within the theme of Environment and Natural Resources, an area of research pursued by several scholars at the IEG in the last decade and more.

The present volume focuses on the Sundarbans, a region in eastern India which is a unique eco-system and World Heritage site, within the larger context of the liberalizing Indian economy. The starting point of the study is the understanding that strategies, which improve or worsen the impacts of resource use on human well-being in a region, often originate outside the region and the relevant sector. In other words, the quantity and quality of services available from a particular ecosystem and economic sectors related to it are largely determined by policies on trade, macro-economy, and a range of other influences originating outside the ecosystem.

In other words, this study adopts a regional focus for analysis of the relationship between ecosystem services and human well-being. Simultaneously, it extends its horizon to the national and international level to identify the causal or driving factors of change. It examines the expansion of aquaculture in the Indian Sundarbans following on export markets generated factors. The conceptual framework that is adopted, therefore links scales of analysis, the economic and ecological. Within such a framework, this study also attempts to study how different dimensions of human well-being might have been impacted. We shall analyse:

i. How the process of export of aquaculture impacts the distribution of income among different stakeholders in the region?
ii. Who are losing and gaining with respect to the different constituents and determinants of human well-being?

Another important feature of the study is its methodological innovativeness. It uses critical inputs from remote-sensing experts, marine scientists, and forestry and fisheries experts and places them in the framework of economic logic and econometric analysis. This provides an understanding of the inter-linkages between biodiversity, land use change, and human well-being in the context of aquaculture in the Sundarbans. It provides deep insights into the nature of the interface between development, well-being of different stakeholders, and conservation of biodiversity, in terms of trade-offs and synergies between components of each. This study shall be of interest to researchers, students, and policy makers looking for policy directions based on inputs from different disciplines.

March 2009

ARJUN SENGUPTA
Chairman, Board of Governors,
Institute of Economic Growth.

Preface

The initial motivation for this study came during our work for the Millennium Ecosystem Assessment (MEA), particularly in our detailed interaction with the Sub-Global Working Group and the multi-scale assessments being carried out by that group and its associates in different parts of the world. It was easy to see that global drivers including trade were indeed changing ecosystems in India, thereby impacting the livelihood and well-being of people dependent on them. We felt the urge to carry out one such study, viewed by us as a multi-scale study of the links between trade, environment, and rural poverty in India. Fast growing aquaculture and the changes it brought about in the Sundarbans region was a good candidate for such a study, in particular due to the unique ecosystem that the Sundarbans represented and its status as a world heritage site.

The opportunity came our way when Anantha Kumar Duraiappah and later Owen Cylke from the Macro Economics Office of the Worldwide Fund for Nature (WWF) in Washington DC visited Delhi in 2003 and our interactions led to the formulation of the India component of their seven-country study on 'Trade, Rural Poverty, and the Environment'. The India component was conceptualized as having a research and an outreach component, the first to be undertaken by the Institute of Economic Growth (IEG) and the outreach by the WWF (India). The research report prepared for that project between 2003 and 2006 forms the basis of the present study. Inputs from two workshops held in Washington DC in 2003 and 2004 and one at Hague in 2005 provided inputs from participants from several organizations as well as from our counterparts in the other countries participating in the project. We are thankful to the Macro Economics Office of WWF (Washington DC) for the financial support and furthermore for the academic interactions resulting therefrom. Grant Milne of the World Bank was very supportive with his comments and suggestions to the study and we extend our thanks to him.

The study was located in West Bengal and our work there was tremendously facilitated by WWF and by Jadavpur University. Gautam Gupta of Jadavpur University facilitated the first pre-project workshop. The School of Oceanographic Studies of Jadavpur University, headed by Sugato Hazra, undertook the analysis of the data on land-use change obtained from National Remote Sensing Agency (NRSA). Abhijit Mitra, of the Department of Marine Sciences of Kolkata University, educated us on the intricacies of prawn seed collection, aquaculture farming, and water pollution. Col. Shakti Banerjee of WWF (West Bengal) facilitated our visits to the interiors of the Sundarbans and Ashish Ghosh's Centre for Development undertook to help with the fieldwork in those areas. Anumitra Anurag Danda assisted in multiple ways in the fieldwork and in the analysis of data from it. Our sincere thanks to all of them. We are also grateful to a large number of people—fishermen, small farmers, and local opinion makers in numerous small islands of Sundarbans who were not only generous with their time and comments but made our stay and visit memorable forever.

Several government agencies provided inputs by way of data and discussions. The Marine Product Export Development Agency (MPEDA) office in Kolkata was very helpful. The Forest Department, in particular A.K. Raha, gave insights and data. Several processing and exporting units were visited by us and we thank them for their patience with our survey questions. Two colleagues from the ISI provided invaluable assistance in the computation of biodiversity indices from time-series data. Madhumita Mukerjee, Deputy Director of Fisheries, Government of West Bengal, and D.P. Jana of the Sundarbans Development Board were also of great help. Our grateful thanks to all of them.

Finally, in the rewriting of the book in its present form, continued interaction with WWF (India) and the ongoing outreach part of the project was of invaluable help. At WWF India, we wish to thank Ravi Singh for his support and Sejal Worah for her enthusiastic participation in numerous discussions, which sharpened our awareness of the limitations of economic analysis.

To sum up, this study extended to far more than was initially thought of. It depended critically on inputs from remote-sensing experts, from marine scientists, and from forestry and fisheries experts. In the final analysis, we have tried to place all the learning in the context of economic logic and econometric analysis to provide an understanding of the inter-linkages between biodiversity, land-use change, and human well-being in the context of aquaculture in the Sundarbans. We hope the reader will find it useful.

Most importantly, we wish to record the contributions of Nisar Ahmed Khan as part of the project team up to 2005. He contributed to the data collection, the field study, and the analysis till he left the IEG. Our thanks also to the staff of the Computer Unit and the Library at IEG, from whom we received constant and competent support and to our colleagues whose comments and suggestions in different seminars and workshops were immensely useful.

We wish to thank the Oxford University Press referee for comments and suggestions which helped improve the analysis and the presentation. Needless to say, the shortcomings, if any, remain our responsibility.

May 2009

Kanchan Chopra
Preeti Kapuria
Pushpam Kumar

Tables, Figures, and Maps

Tables

1.1	The Ecological Footprint of Seafood Production	11
1.2	Volume and Value of Aquaculture Production at a Glance	12
2.1	Initiatives Taken by the DFFARFH	25
2.2	Yearly Shrimp Production by Major Producing Countries 1991–2000	28
2.3	State-wise Shrimp Production in India	30
2.4	Share of Marine Products Export in Agricultural Exports and Total Exports of India	31
3.1	Export of Frozen Shrimp from West Bengal (Random Effects Model)	54
3.2a	Export of Frozen Shrimp from West Bengal (Classical Regression Model)	56
3.2b	Export of Frozen Shrimp from West Bengal (Classical Regression Model)	57
4.1	Land Converted to Shrimp Culture in the Sundarbans—1986 to 2004	96
5.1	Cost Shares of Inputs in Aquaculture	119
5.2	Estimates of Translog Total Cost Function	120
5.3	Own Elasticity of Substitution (Allen)	121
5.4	Own Price Elasticity of Demand	121
5.5	Cross-price Elasticities	122
5.6	Cross Elasticity of Substitution (Allen)	123
5.7	Estimates of Translog Total Cost Function	125

List of Tables, Figures, Maps and Appendices   xv

6.1	Block-wise Total Land Transformed to and from Aquaculture and Total Land under Aquaculture during 1986–2004	145
6.2	Block-wise Trend in Land-use Transformations from Various Classes to Aquaculture during 1986–2004	146
6.3	Conversion of Paddy Land to Aquaculture (Random Effects Model)	155
6.4	Conversion of Mangrove Land to Aquaculture	157
7.1	Sen's and Nussbaum's Approach to Capability	172
7.2	Block-wise Average Income of a Shrimp Farmer	178
7.3	Block-wise Average Income of a Shrimp Farm Worker	179
7.4	Block-wise Average Income of a PL Collector	179
7.5	Income and Employment from Shrimp Production and Processing for Export in the Indian Sundarbans	180
7.6	Block-wise Average Income of an Agricultural Farmer	181
7.7	Measurement of Indices of Well-being/Ill-being	182
7.8	Indicators of Well-being of Agricultural Farmers in Canning and Minakhan	184
7.9	Indicators of Well-being of Shrimp Farmers in Canning and Minakhan	185
7.10	Indicators of Well-being of Shrimp Farmers and Mixed Income Households in Canning and Minakhan	187
7.11	Indicators of Well-being of PL Collectors and Fishermen in Gosaba	189
7.12	Indicators of Well-being of PL Collectors and Salary/Wage Earner in Gosaba	191

Figures

2.1	Export Growth of Indian Marine Products (Volume)	32
2.2	Export Growth of Indian Marine Products (Value)	32
2.3	Composition of Marine Products Exports from India (Volume)	33
2.4	Composition of Marine Products Exports from India (Value)	34
2.5	Destinations of Marine Products Exports from India (Value)	35–36

2.6	Percentage Distribution of West Bengal in the Total Frozen Shrimp Exports from India	37
3.1	Composition of Marine Products Exports from West Bengal (Value)	47
3.2	Major Destinations of Frozen Shrimp Exports (Value) from West Bengal	48–49
4.1	Different Stakeholders in Aquaculture	93
7.1	Ecosystem Services and Human Well-being	175
7.2	Indicators of Well-being of Shrimp Farmers and Agricultural Farmers in Canning and Minakhan	186
7.3	Indicators of Well-being of Shrimp Farmers and Mixed Income Households in Canning and Minakhan	188
7.4	Indicators of Well-being of PL Collectors and Fishermen in Gosaba	190
7.5	Indicators of Well-being of PL Collectors and Salary/Wage Earner in Gosaba	191

Maps

1.1	Sundarbans: Blocks, Forest Divisions, and the Tiger Reserve	17
3.1	Location of West Bengal in India	46
4.1	Sundarbans in West Bengal	87
6.1	Land-use Map of Sandeshkhali in 2004	147
6.2	Land-use Map of Minakhan Block in 2004	148
6.3	Land-use Map of Namkhana in 2004	148
6.4	Land-use Map of Basanti in 2004	149
6.5	Land-use Map of Canning I & II in 2004	149
6.6	Land-use Map of Kakdwip 2004	150
6.7	Land-use Map of Gosaba 2004	150
6.8	Land-use Map of Kultali in 2004	151

ABBREVIATIONS

AES	Allen Elasticity of Substitution
ARTCA	Antimicrobial Regulation Technical Corrections Act
BFFDA	Brackish Water Fish Farmer's Development Agency
BIS	Bureau of Indian Standards
CAC	Codex Alimentarius Commission
CCP	Critical Control Point
CGR	Compound Growth Rate
CIFRI	Central Inland Fisheries Research Institute of India
COC	Code of Conduct
CPR	Competitor's Price
DFAARFH	Department of Fisheries, Aquaculture, Aquatic Resources, and Fishing Harvest
DO	Dissolved Oxygen
EC	European Commission
EIA	Export Inspection Agency
EIC	Export Inspection Council
ETP	Effluent Treatment Plant
EU	European Union
FAO	Food and Agriculture Organisation
FDA	Food and Drug Administration
FSS	food safety standards
GMP	Good Manufacturing Practice
GS	General Stringency
HACCP	Hazard Analysis Critical Control Point
HDI	Human Development Index
HHS	Health and Human Services

IEG	Institute of Economic Growth
IFS	International Financial Statistics
ISODATA	Interactive Self-Operational Data Analysis Technique
L&P	Labelling and Packaging
LERM	Liberalized Exchange Rate Management System
LM	Lagrangian Multiplier
MEA	Millennium Ecosystem Assessment
mha	million hectare
MOHFW	Ministry of Health and Family Welfare
MPEDA	Marine Product Export Development Agency
MRL	Maximum Residual Limit
NCCP	National Codex Contact Point
NRSA	National Remote Sensing Agency
NSSO	National Sample Survey Organisation
NTM	non-tariff measure
PFA	Prevention of Food Adulteration Act
PHSA	Public Health Service Act
PL	Post-larvae
PPM	Product and Process Method
RP	Relative Price
SC	Scheduled Caste
SCE	Scale Economies
SPS	Sanitary and Phyto-sanitary
TBT	Technical Barriers to Trade
TC	Total Cost
UNDP	United Nations Development Programme
WCA	Water Cess Act
WHO	World Health Organization
WPSI	Wildlife Protection Society of India
WTO	World Trade Organisation
WWF	Worldwide Fund for Nature

1

Ecosystems, Ecosystem Services, and Aquaculture
Drivers of Change

SETTING THE STAGE: NATURAL CAPITAL, RESOURCE USE BY THE ECONOMIC SYSTEM, AND ECOSYSTEM SERVICES

Achieving higher levels of human well-being has been the motivating force behind changes in the use of natural resources such as land and water over centuries of human endeavour. The institutions that define our economic systems and their functioning aim at improving the lot of humans. Different stages in this story of human achievement can be said to be marked by the changing use of land: initially through settled agriculture, and subsequently by adoption of better technologies for harnessing water for irrigation, and for better modes of cultivation. Later developments move on to industrialization and urbanization with more complex and indirect linkages with natural resource use.

Simultaneously, natural scientists view natural resources as constituents of ecosystems, which are 'a dynamic complex of plant, animal, and micro-organism communities and their non-living environment interacting as a functional unit'. Ecosystems may be of many kinds, dominant land-use determining the category into which they fall.[1] Human beings, one component of this functional unit, through the modes of their resource use, impact the direction and variability of changes, which take place within these functional units over time. They use these ecosystems as a kind of 'natural capital', which augments society's productive base together with manufactured capital, human capital, and social capital.

As expounded by Dasgupta and Maler (1994), 'An economy's productive base consists of its capital assets and its institutions. Economists have recently shown that the correct measure of that base is wealth. More precisely, they have shown that in estimating wealth, not only is the value of manufactured assets to be included (buildings, machinery, and roads), but also 'human' capital (knowledge, skills, and health), natural capital (ecosystems, minerals, and fossil fuels), and institutions (government, civil society, and the rule of law)'.

Such an approach to viewing the productive base of an economy has led to two strands of literature. One strand maintains that development is sustainable so long as an economy's wealth relative to its population is maintained over time and that economic growth should be viewed as growth in wealth, and not growth in GNP. Several studies have followed this line of thinking in recent years to estimate 'real wealth' and 'real savings'.[2] The recent World Bank study (2006b) mentioned here refers to 'the millennium wealth assessment' and has laid claim to moving towards assessing the wealth of the planet in the year 2000.

The second strand of literature argues that 'natural capital' is special in a number of ways. First, some kinds of natural resources cannot be reproduced. Other natural resources are reproduced by processes not entirely within the capability of human beings to control, for example, rainfall and temperature. So, changes in natural capital are neither easy to measure nor easily controlled, at least at macro-levels. Additivity across large spatial entities of different kinds of natural capital is difficult, if not impossible. How then do we know whether the use we are making of ecosystem-based resources is sustainable and not an over-use, which will erode our own well-being?

Further, some stocks of natural capital produce incomes and utilities by virtue of their existence, without the intermediation of a production process.[3]

It is easier to track and understand the behaviour of ecosystems within which natural capital is embedded at less aggregated meso-levels of analysis, say a coastal or a forest ecosystem. Is it not better then to focus on scales of analysis amenable to an understanding of the relationship between services provided by ecosystems, the driving forces leading to changes therein, and the dimensions of well-being they impact? Such a focus enables us to understand the multi-dimensionality of the ecosystem services, can take into account the many ways in which stocks of capital contribute to the many dimensions of human well-being.

Simultaneously, important drivers (both economic and ecological) operating across scales of analysis can be identified to help in deriving policy conclusions. *This study adopts a regional focus for analysis of the relationship between ecosystem services and human well-being. It also extends its horizon to the national and international level to identify the causal or driving factors of change. It examines the expansion of aquaculture in the Indian Sundarbans following on export markets-generated factors. The conceptual framework that we adopt therefore, links scales of analysis, the economic and ecological.*

The starting point of the study is the understanding that strategies, which improve or worsen the impacts of resource use on human well-being in a region, often originate outside the region and the relevant sector. In other words, the quantity and quality of services available from a particular ecosystem and economic sectors related to it are largely determined by policies on trade, macro-economy, and a range of other influences originating outside the ecosystem.

Such a cross-scale, regional analysis is also perhaps the only methodological approach to dealing with the problem of the '*misfit between societal institutions and ecosystems*'.[4] Economic and social institutions are formal and informal constraints (North 1990) and their enforcement characteristics shape incentives in economic exchange. They are often not as spatially differentiated as ecosystems. Driving forces get transmitted to ecosystems with varying degrees of efficiency, without taking into account the spatial variations intrinsic in ecosystems. They may, therefore, have differential impacts, depending on the state of the ecosystem and its resilience.[5] Some kinds of change or disturbance may trigger a movement towards a better equilibrium; in other cases the local system, if it is a fragile one, may be overwhelmed with shocks from the external world. Studies, which dwell on the impact of global or national drivers on local systems, assist in unravelling the pathways through which economic drivers impact ecosystems. In this volume we intend to adopt just such an approach in studying aquaculture driven by international prices and its impact on the Indian Sundarbans. In particular, the impacts of aquaculture expansion on biodiversity and on land-use shall be examined in depth.

SETTING THE STAGE: APPROACHES TO HUMAN WELL-BEING

Economic systems do not change in a vacuum. Most drivers of change are expected to promote human well-being and improve the human condition. This is particularly important when we are dealing with a region in eastern India, which is characterized by considerable poverty and ill-being. However, policy makers in countries across the world often consider monetary income or per capita GDP as a measure of human well-being. The literature has of course evolved in several directions from that simplistic understanding.

Poverty is defined as 'the pronounced deprivation of well-being' in the *World Development Report* (2001b). This definition of poverty leads to the question: What then is well-being? The literature on well-being accepts its multi-dimensional characteristics and the *World Development Report* takes this multi-dimensionality as its reference point.

It is appropriate to mention here that this approach is related to, though distinct from, the 'capabilities' approach of Amartya Sen. In the latter, development is defined as *an expansion of capabilities*, leaving the selection of the relevant capabilities as a value judgment. Sen refrains from developing a list of basic capabilities or even setting up a procedure for doing so. This constitutes both a strength and a weakness of his capability approach. While it helps to avoid the problems that may emerge from an over-specification of human nature, it is open to the criticism of not being easily amenable to operationalization.[6]

Development practitioners in particular see a great deal of use in drawing up lists that incorporate the different constituents and determinants of well-being. Such specifications help governments in policy direction and in measuring performance against given directions of change. They also contribute towards arresting the tendency of policy makers to couch themselves in a framework where utility (state of deriving satisfaction) is primarily and predominantly dependent upon income and subsequent consumption. Empirical work in a large number of countries has been of use in assisting to see what values humans hold dear and in therefore, identifying what may alternately be called 'basic capabilities to be striven for' or 'constituents and determinants of well-being'.[7]

Narayan et al. (2000a) provided a list based on evidence from a large number of developing countries. They came up with a list of constituents and determinants of well-being, which was examined along with others in a recent conceptual exercise carried out by the Millennium Ecosystem Assessment (MEA).[8] Among other things, the MEA was charged with the task of examining the literature on human well-being in order to determine its links with ecosystem services. The overwhelming evidence from the studies reviewed is that poor people's idea of a good life is multi-dimensional. The dimensions cluster around the following themes: material well-being, physical well-being, social well-being, security, and freedom of choice or action. Communities scattered across different countries also included a sense of 'responsible well-being' in their understanding of what constituted good life.[9]

Taking into account both the state of theoretical developments and empirical evidence documented, the MEA identified the constituents and determinants of human well-being as follows:

1. Basic material for a good life (adequate livelihoods, sufficient food, shelter, access to goods),
2. Health (strength, feeling well, access to fresh air, and water),
3. Social relations (social cohesion, mutual respect, ability to help others),
4. Security (personal safety, secure resource access, security from disasters), and
5. Freedom and choice (opportunity to be able to achieve what an individual values; doing and being).

However, as has been well documented, even in the early literature, 'a major problem is that historically growth has expanded choice in some directions while constricting it in others'.[10] Such a situation might well have come about in the ecosystem of the Sundarbans. This study attempts to show how different dimensions of well-being might have been impacted. We shall analyse:

i. How the process of export of aquaculture impacts the distribution of income among different stakeholders in the region and
ii. Who are the losers and gainers with respect to the different constituents and determinants of human well-being listed above.

MAJOR DRIVERS OF GROWTH IN AQUACULTURE IN THE WORLD ECONOMY

Traditionally grown aquaculture including fish and shrimp has, over the years, given way to more stable production with precision inputs. Intensive and semi-intensive form of cultivation, which was characterized by stocking density, became pervasive. Between 1980s and 1990s the shrimp aquaculture has expanded by a factor of seven (Rosamond et al. 1998). Some of the major drivers of aquaculture growth are:

(i) *Expanding market*, the systematized retail shops, and departmental stores in Europe, Japan, North America, and south-east Asia created the need for stable supply of aquaculture products at a large scale,

(ii) *Adaptation and innovation*, shrimp production has benefited from technological innovation of the 1970s. Farmers living in coastal regions of tropical countries have adapted to shrimp farming quite well,

(iii) *Price competition*, growing price of captured fish and shrimp created the need for greater production from cultured method,

(iv) *Changing consumer preferences*, worldwide there has been a change in food preferences, lifestyle, and work force participation, especially by women. All these new situations also created a niche for aquaculture products,

(v) *New emerging markets*, not only in Europe, North America, and Japan but also in entire South-east Asia and North Central Africa, the market for aquaculture has flourished significantly for the last two decades,

(vi) *Productivity gains in a dynamic global market*, this has created favourable conditions for aquaculture,

(vii) *Emergence of trading blocks like European Commission (EC), NAFTA, ASEAN etc.,* have removed the existing physical, fiscal, and technical barriers to trade. This created appropriate enabling conditions for aquaculture trade, and

(viii) *Technological innovation in production, improved processing and warehouse facilities, E-commerce and positive intervention by the national government, and multilateral organizations like the World Bank and Asian Development Bank.*

It is not surprising that the global trade in fish and fisheries products has grown from US$ 15 billion in 1980 to US$ 64 billion in 2003. Developing countries account for more than 48 per cent (US$ 30 billion) of the global exports. In future too, the growth is likely to continue as the pace of population growth, incomes, and urbanization in the developing countries grow and the stock of wild fish approaches its biological and physical limits. Some projections suggest that in three different scenarios of stagnating captured fisheries production, aquaculture output is required to grow between 1.4 to 5.3 per cent per year in order to bridge the projected future supply gap and provide in the order of 70 million tons of food fish by 2020 (World Bank, 2006a).

ENVIRONMENTAL IMPACTS OF AQUACULTURE

While the growth of aquaculture has been noted as a welcome move, its environmental impacts are also becoming the cause of concern from various quarters. Income and employment generation have been the hallmarks behind this upbeat mood over aquaculture growth. On the other hand, its seed collection, farming, and processing are resulting in coastal water pollution, loss of biodiversity, and habitat fragmentation, which are proving to be matters of concern for policy planners and civil society. These ecological impacts of aquaculture seriously limit the long-run sustainability of this activity. These issues are contentious and may imply significant trade-offs and hence value judgement-based policy implications. In this section, we attempt to list the entire gamut of environmental impacts of aquaculture on bio-geophysical environment. The added focus will be on land-use change and biodiversity loss in the Indian Sundarbans.

Under shrimp aquaculture, young shrimp, usually tiger prawn shrimp (*Penausus monodon*) and pacific white shrimp *(P.vannameis)* are cultured with the help of fertilizers, pesticides, and other nutrients. Asian coastal aquaculture is dominated by seaweed, mollusce, and shrimp and the pond production is dominated by *p.monodons* and *p.merguensis*. In fact, these two varieties of shrimp through intensification and extensification touched the production figure of 0.65 million tons in 1994 from 0.1 million tons in 1984 (FAO 1996). These shrimps are reared till they reach the marketable size by aeration of water and use of chemicals. In this entire process, the aquaculture activities inflict following types of impact on the surrounding bio-geo-physical environment:

1. Degradation of habitat such as mangrove systems,
2. Salinization of soil and water,
3. Coastal and freshwater pollution,
4. Alteration of local food web and ecology,
5. Depletion of wild resources and biodiversity for seed,
6. Depletion of wild resources like fish variety (caused by prawn seed collection), and
7. Impact of introduction of species.

Although there are several benign impacts of aquaculture as well, like water treatment, weed control, nutrient sink etc., but they are comprehensively outweighed by the costs inflicted through the malign impacts (Folke et al. 1998).

Amongst the several ecological damages emanating from aquaculture, loss of fragile ecosystems of mangroves especially in the regions of the Indian Sundarbans are most alarming. Mangroves contribute to societal well-being in numerous ways. They are critical for the coastal communities as they provide major types of ecosystem services—provisioning, regulating, cultural, and supporting (MEA 2003). Coastal communities get livelihood sources like fish, honey, forage, medicines, and various other food items. Other services include necessary breeding ground for fish, crabs, shrimp, buffers against storm surge and shoreline erosion, absorption of pollution, maintenance of biodiversity, and conservation of water. In fact, shrimp cultivation in coastal areas is possible because of the presence of mangroves as the necessary seed, shrimp ponds, and food inputs, that come from this fragile ecosystem. Aquaculture is also protected from natural hazards by mangroves. Through various life-supporting services, mangroves provide sustainability to aquaculture. One can well imagine the amount of costs the farmers would have to incur if mangroves are non-existent. In recent years, the unsustainable land-use policies and myopic policy choice has done the maximum— damage to mangrove forest all over the world. Today 50 per cent of the mangroves' ecosystems are already converted or destroyed by humans (Rosamond et al. 1998). The aquaculture-driven conversion of mangrove has been the major reason. It is estimated that more than 60 per cent of Asia's mangroves have already been converted to aquaculture farms, primarily for the production of shrimp. Asia has the greatest concentration (41.5 per cent) of the world's 18 million hectares of mangroves, and South-east Asia the widest expanse (at least one million ha) of brackishwater aquaculture ponds with largest expanse of 4.5 million ha in Indonesia (Primavera 2000). The incremental costs of mangrove conversion to shrimp ponds is quite high. And these costs remain unacknowledged, unaccounted, and un-factored for into the decision-making processes.

ECOLOGICAL FOOTPRINT OF AQUACULTURE

The Modern form of aquaculture done in intensive and semi-intensive ways with considerably high stocking density, is known to have profound impacts on the environment. One of the major impacts happens to be the conversion of precious agricultural land and mangrove land into aquaculture farmland. Usually there is conversion of agricultural field and land adjoining the mangroves, which are ecologically fragile. One of the serious lacunas of modern aquaculture is that it is driven by current revenue maximization and hardly pays any attention to long-term ecological balance (Folke et al. 1998, Gunawardena and Rowan 2005). Internalizing these ecological costs into the pricing structure would be a possible policy response but not very easy considering the existing political structure in those countries, prevailing terms of trade, and the world economic order. However, internalization of these ecological costs would reveal the costs, which the society (invariably poor people in the aquaculture exporting country) is paying for its consumption and preferences. Ecological costs, if embedded into the pricing, would also convey the externality borne by developing countries.

Activities, which are natural resource-intensive like aquaculture, have serious ecological implications which impact society and human well-being. Aquaculture, having an impact on the state of ecosystem, impairs the ability of the ecosystem to perform its functions, which have beneficial values for the society. Modern aquaculture seems to emerge as one such activity, especially in coastal areas and in the vicinity of mangroves. This can be better understood with the help of the concept of ecological footprint. Rees and Wackernagel (1994) explain ecological footprint as the land area necessary to sustain current levels of resource consumption and waste discharge by a human population. They were the first to introduce this concept but the spirit of the concept goes back to Bogstrom's 'ghost acreage' reflecting areas of agricultural land required for fuel consumption and Odum's (1985) 'energy' showing the amount of energy consumed per unit of area per year. Rees and Wackernagel estimated that the Fraser Valley, Vancouver depends on an area 19 times larger than that contained within its boundaries, for food, forestry products, carbon dioxide assimilation, and energy. They go further and suggest that it would not be possible to sustain the present human population of more than six billion people at the same material standard as that of the US without having at least resources of two additional planets (Rees and Wackernagel 1994). In this context, sometimes, another concept that is 'carrying capacity' is also used and it is defined as the maximum rate of resource consumption and waste discharge that can be sustained indefinitely without progressively impairing the functional integrity and productivity of ecosystems.

Some commentators maintain that ecological footprint by all means is a static concept. Ecosystems are dynamic and are characterized by a complex nature with presence of non-linearity, thresholds, and discontinuity (Costanza et al. 1993). Ecological footprint may not be able to capture the dynamic aspect of ecosystems and ever changing equilibria but it does shed some light on the precise requirement of human activity like modern aquaculture.

Ever-expanding aquaculture is projected as a saviour of growth and prosperity in developing countries but monoculture-dominated aquaculture uses ecosystems services for the purposes of culturing. It uses ecosystem for its all inputs requirements—feed, seed, water, waste treatment etc. Folke et al. (1998) have estimated the ecological footprint of seafood production.

For shrimp pond farming, the requirement is 34–187 hectares per hectares of the farming area. Waste assimilation also needs 2–22 ha / ha of farming. Folke et al. 1998 go on to suggest that the implication of the size of the supporting mangrove nursery area becomes clearer when shrimp farming is analysed at a national and regional level where usually the mangrove nursery area for post-larvae extends far beyond the physical location of the shrimp farms.

Table 1.1: The Ecological Footprint of Seafood Production
(Values are area of footprint per area of activity, ha/ha)

Activity	Resource Production Support	Waste Assimilation Support
Salmon cage-farming, Sweden	40,000–50,000	–
Tilpia cage-farming, Zimbabwe	10,000	115–275
Fish tank system, Chile	–	16–180
Shrimp pond farming, Columbia	34–187	–
Shrimp pond farming, Asia	–	2–22
Mussel rearing, Sweden	20	–
Tilpia pond farming, Zimbabwe	0	0
Cities in the Baltic Sea Drainage basin	133	–

Source: Adapted from Folke et al. 1998.

Thus, they are contrary to the idea of sustainable practice of aquaculture farming. In Sundarbans, the way the prawn seeds are collected by the locals causes serious damage to the wild fish and other coastal organisms. Aquaculture in the region remains largely dependent upon wild caught seed. This is likely to have serious consequences for the coastal biodiversity.

TRADE IN SHRIMP AQUACULTURE: GLOBAL AND NATIONAL SCENARIO

Shrimp along with salmon constitutes the major share in aquaculture in terms of value and volume of global trade. Aquaculture as a whole has experienced an added momentum in production and trade all over the world in the last three decades (1975–2005). The growth has primarily been in the developing countries during 1985–2005. Aquaculture is farming of aquatic organisms like fish, shrimps, crustaceans, and many other species for food and ornamental purposes (for example, pearl). The most distinctive feature of aquaculture is its controlled production with greater precision in inputs. The FAO defines aquaculture as 'the farming of aquatic organisms in inland and coastal areas involving interactions in the rearing process to enhance production and the individual or corporate ownership of the stock being cultivated'. Usually aquaculture is used to refer to fish, shrimp, and other aquatic species. The International Standard Industrial Classification of All Economic Activities recognizes aquaculture as a separate activity—

Table 1.2: Volume and Value of Aquaculture Production at a Glance

Country	By M tons	Quantity per cent	By $ million	Value per cent	$'000 tonne
China	30.6	67.3	30.870	48.7	1.01
India	2.5	5.4	2,936	4.6	1.19
Vietnam	1.2	2.6	2,444	3.9	2.04
Thailand	1.2	2.6	1,587	2.5	1.35
Indonesia	1.0	2.3	1,993	3.1	1.91
Bangladesh	0.9	2.0	1,363	2.2	1.49
Japan	0.8	1.7	3,205	5.1	4.13
Chile	0.7	1.5	2,801	4.4	4.15
Norway	0.6	1.4	1,688	2.	2.65
USA	0.6	1.3	907	1.4	1.50

Source: The World Bank 2006a.

although for recent years only the data on aquaculture is provided separately from the data on fisheries.

Traditionally farmers in tropical climate located near the fresh and marine water have been growing shrimp and other species for subsistence consumption. Since the 1980s production has picked up and trade has accelerated. The average rate of growth of aquaculture has been more than 10 per cent per annum. Since the 1980s it reached 259.4 million tons with the value of, US$ 70.3 billion in 2004.

It is also noteworthy that captured fisheries have grown hardly at the rate of 2 per cent per annum for the same period. Although the aquaculture has obtained the status of a global industry, the share of developing countries is more than 90 per cent. Out of this Asian countries contribute to 89 per cent of aquatic production (80 per cent in value terms) (World Bank 2006a). China has a lion share at 67 per cent and 49 per cent in volume and value terms respectively among Asian nations followed by India.

Indian aquaculture has experienced a six and half-fold growth over the last two decades, with freshwater aquaculture contributing over 95 per cent of the total aquaculture production. Carp production mainly dominates freshwater aquaculture whereas shrimps are grown in brackish water. The three Indian major carps, namely catla (*Catla catla*), rohu (*Labeo rohita*), and mrigal (*Cirrhinus mrigala*) contribute the bulk of production with over 1.8 million tons (FAO 2003); followed by silver carp, grass carp, and common carp forming a second important group. Average national production from still water ponds has increased from 0.6 tons/ha/year in 1974 to 2.2 tons/ha/year by 2001–2.

While carp and other finfishes are grown for the domestic market, a large proportion of freshwater prawn production is exported. In contrast, the development of brackish water aquaculture has been confined to a single species, *Penaeus monodon*, the scientific farming of which began only recently during the early 1990s. The area devoted to shrimp farming extends to as much as 152 thousand ha producing approximately 115 thousand tons, the majority of which is destined for export. During 2002–3, cultured shrimp and prawn contributed 65.7 per cent of their total exports, mainly in frozen form and with a value of over US$ 0.80 billion. Rough estimates put the share of aquaculture in Indian fish and fishery products exports to 22 and 55 per cent in volume and value terms respectively.

Trade Regulations Concerning Aquaculture Industry

With increasing volume of production and trade, global awareness of and concerns over environmental issues related to aquaculture have increased. Fish products in general and aquaculture products in particular have been subject to close scrutiny for their safety and environmental impacts. There is an increasing demand for improved environmental sustainability and human health safety from the aquaculture sector. Furthermore, food safety standards have become stringent and the international trade regulations tightened. Although, traditional barriers to trade such as tariffs and quantitative restrictions are at least partially liberalized and many developing countries are subject to preferential trading arrangements, other measures such as food safety requirements can equally act as barriers to trade (Spencer Henson et al. 2000). The technical regulations and standards to fisheries products are in the form of non-tariff measures (NTM) that include, besides others, Sanitary and Phyto-sanitary Measures (SPS), labelling, and certification requirements. The most important food safety standard is the agreement on the application of SPS measures, which applies only to measures covering food safety, animal and plant life, and human health. Other technical measures outside this area come within the scope of the Technical Barrier to Trade (TBT) Agreement. All developed countries and a large number of developing countries are required to take up regulatory Hazard Analysis Critical Control Point (HACCP) systems. Critical Control Point (CCP) means a step or procedure with which control can be applied and health hazards can be prevented, eliminated, or reduced to an acceptable level. In applying HACCP, all food-borne hazards like Salmonella, E.Coli, Listeria monocybgenes, etc., are to be considered. An essential prerequisite to HACCP, therefore, is the adoption of Good Manufacturing Practice (GMP). Fish and fish products from aquaculture are included, either explicitly or implicitly, in such HACCP-based regulations.

The provisions concerning the use of HACCP systems require significant investment. The investment requirements for HACCP plants are huge, as most of the capital goods related to the plant are to be bought from the developed countries. The cost of complying with the food safety standards, laid out by the developed importing countries, imposes a heavy burden on the developing countries, making them uncompetitive in the export market. Anjani Kumar and Praduman Kumar (2003) explored the opportunities of trade liberalization and challenges put forward by the SPS measures and TBT agreements with special reference to India's fisheries' exports. One of the main findings of the study was that the compliance with food safety measures is a costly proposition for developing countries like India as it has an adverse affect on its export competitiveness. However, the authors concluded that food and safety concerns are vital and the exporting countries have to comply with to promote exports. Besides high cost of compliance, issues like irrelevance of foreign standards to local conditions, lack of timely and adequate information and high transaction costs, difficulties in understanding requirements, changing quality standards of the importing countries as well as testing and monitoring them are some of the major problems faced by the processing/exports units.

Mehta and George (2005) have observed that information regarding procedural norms and regulations of various countries (for example, sampling, inspection, and test) is totally lacking. The authors found out that the stringent regulations prevailing in the developed countries lead to higher compliance costs besides restricting the exports of processed food to these countries. According to them, India stood to lose because of inter-country differences in the food safety standards (FSS). In addition, there are evidences to prove that Indian processed food exporters face discriminatory treatments in the application of food safety standards especially in the European Union, United Kingdom, and the United States.

Shyam et al. (2004) have studied the export performance and potential of Indian marine products under the trade liberalized economy and explored the possible impacts of WTO agreements on the Indian fisheries sector. The study examines the growth, instability, structure and direction of export, potentials, and opportunities in order to ascertain the problems and impediments faced by the exporters of the marine products. The analysis is based on primary data collected from 30 seafood exporters in India. The study concludes that the competitiveness of major marine products *except for shrimp* has decreased during the post-liberalization period when compared with the pre-liberalization period. The rejection from the European Union on account of the microbial, antibiotic and bacterial residues, quality issues, and higher domestic demand threatens the competitiveness of squid, cuttlefish, and pomfrets. High cost of investment, scarcity of raw material, dictatorship of the buyers, and the low capacity utilization are the major problems encountered in the fisheries export.

THE SUNDARBANS: A BIODIVERSE, FRAGILE, AND CHANGING ECOSYSTEM

The Sundarbans is a region where the biodiversity is rich and valued. The tiger reserve comprising 2,585 square kilometers of the Sundarbans National Park and its buffer zone is a part of this region. The national park was declared a UNESCO world heritage site in 1989. Two wildlife sanctuaries are also located within the region. A total of 69 floral species (included within 29 families and 50 genera) have been recognized in the Sundarbans area at the north-eastern coast of the Indian subcontinent, out of which 34 species are true mangrove types. This ecosystem sustains almost all the mangrove species available in other parts of the sub-continent, Burma, and other South-east Asian countries. However, due to habitat destruction, human interferences, and salinity fluctuations, this ecosystem is presently under great stress. The Sundarbans ecosystem is still one of the most biologically productive and taxonomically diverse ecosystems of the sub-continent, although about six vertebrates have disappeared from this ecosystem since the last 200 years and about 20 species are in the endangered species list.

The Forest management divides the area into (a) core zones, (b) buffer zones, and (c) manipulation zones, which are made up of forestry, agriculture, and aquaculture zones. The different areas support each other and in turn provide ecosystem services to the people of the region. Nutrient supply for instance comes from the mangrove forests. Salinity of water decreases landward within the rivers so that paddy and other agricultural cultivation is carried on

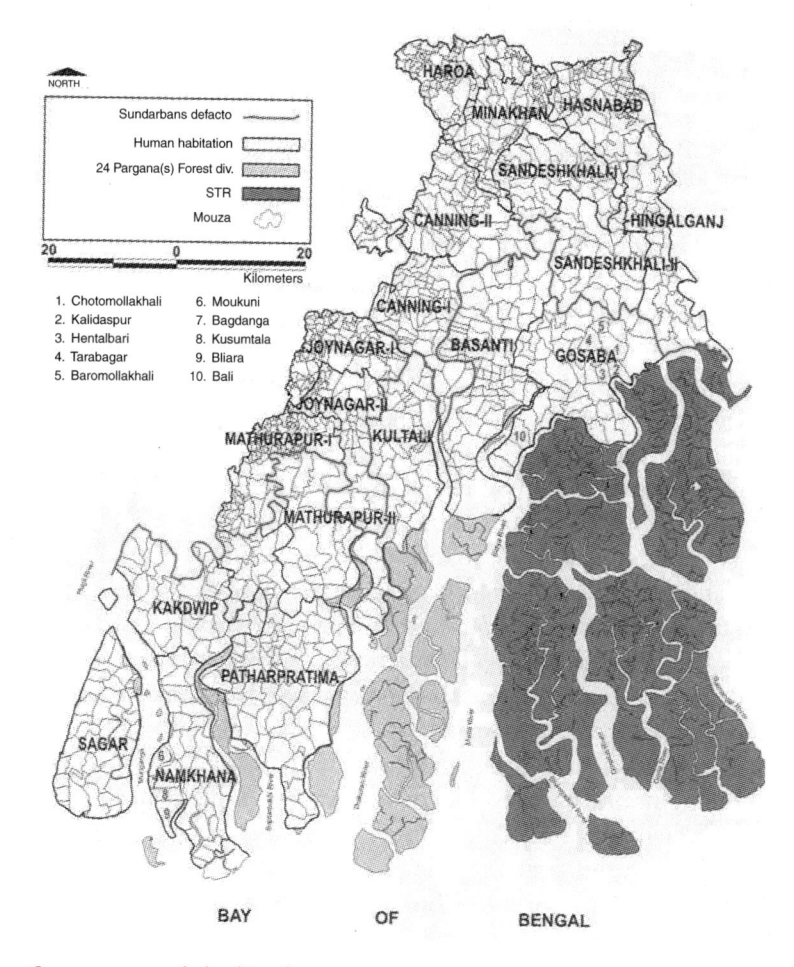

Map 1.1: Sundarbans: Blocks, Forest Divisions, and the Tiger Reserve

there. The presence of mangroves near the coast provides important storm protecting and other regulating services. In other words, a variety of eco-system services falling within the groups of provisioning, regulating, and cultural accrue simultaneously from this ecosystem.

The mangrove forests[11] in the region are home to a variety of marine and terrestrial organisms. Many indigenous coastal communities rely on mangroves to sustain their traditional cultures. A large portion of the population is dependent on the forests for fishing, prawn seed collection, fuelwood, honey collection and other non-wood forest products, wood for making boats, construction materials for thatching houses, and medicinal plants. Fishing is the second largest occupation after agriculture for the coastal communities. The majority of people in the Sundarbans region are crop farmers. Most of them also operate shrimp farms and others raise carp, sea bass, and freshwater prawn in ponds, and in small water storage reservoirs. A large percentage of the rural population is landless or own small farms. The landless rural poor work as daily labourers on aquaculture farms, agricultural fields or find jobs in fish markets. Some of them also travel to nearby cities in search of jobs. Some of the very poor rural households are also involved in prawn seed collection from the wild.

The increasing human population in the region, largely due to migration from Bangladesh, has led to a large-scale conversion of the mangrove forests to aquaculture ponds and agricultural fields. This, coupled with the regular practice of collection of tiger prawn seeds by dragging the nylon nets along the riverbanks, has adversely affected the mangrove ecosystem of the region.

The region is also subject to a series of changes from natural causes including rise in sea levels due to temperature changes in the long run. The temperature is expected to rise at an average of 0.19 degrees per year C in this region. Sea ingression has been a feature of this area over the past three to four hundred years and the rate may rise in the future.[12] As a consequence, land is an extremely scarce resource. A multiplicity of causes makes the region a fragile and vulnerable natural system. Assessing the biophysical impacts of an increase in aquaculture in the region is indeed a major challenge. To add to this, migration of people to areas of increased aquaculture concentration has added to its vulnerability.[13]

STRUCTURE OF THE BOOK: A PREVIEW

In this volume, we propose to examine the impact of increased aquaculture on biodiversity, land-use, and human well-being in the Indian Sundarbans. Chapters 2 and 3 examine the drivers of shrimp export and the issues faced by exporting concerns. Chapter 4 describes the region and the stakeholders in the value chain created by the increased export of shrimp. Chapters 5 and 6 use data and methodology of diverse kinds to determine the extent of biodiversity loss off the Sundarbans coast and the land-use change in the region. Chapter 7 quantifies the impact on well-being of the different categories of stakeholders. The final chapter sums up and provides pointers towards possible policy recommendations for sustainable aquaculture in the region.

NOTES

1. The Millennnium Ecosystem Assessment (2005) identifies nine kinds of ecosystems, main among them being forest, marine, agricultural, dryland, and fresh water.
2. See, for instance, World Bank (2006b), Hamilton and Clemens (1999), Hamilton and Hartwick (2005), and earlier Dasgupta and Maler (1994).
3. Utility and incomes accrue, for instance, from a 'stock' of aesthetic landscapes, even by its very existence and without depreciating a part of the capital.
4. See Pritchard et al. (1998).
5. Resilience is the system's capacity to survive disturbance. See Berkes, Colding, and Folke (2003) for greater details on approaches and implications.
6. Sugden (1993) gives one such critique. Stewart (1996), Gasper (1996) are others who advocate movement towards a basic set of capabilities. Nussbaum (2000) has in fact, drawn up a list of central human capabilities.
7. See Dasgupta (2001) in this context.
8. See Millennium Ecosystem Assessment (2003), in particular Chapter 3 on Human Well-being.
9. This led Chambers (1997) to consider 'responsible well-being' as the main agenda for development.
10. See Appffel-Marglin and Marglin (1990).
11. A mangrove is a woody plant or plant community, existing between the sea and the land in areas, which are inundated by tides. Thus a mangrove is a species as well as a community of plants. Mangroves have immense ecological value, being the second highest source of primary production next to rainforests. They protect and stabilize

the coastal zone, nourish the coastal waters with nutrients, yield commercial forest products, support coastal fisheries, and provide many of the resources upon which coastal people depend for their survival and livelihood. This unique coastal ecosystem sustains a rich floral and faunal community in and around its vicinity besides providing various direct and indirect benefits to the stakeholders.

12. According to a study of the School for Oceanographic Studies, Jadavpur University.
13. According to the study of Jayshree Roy Chaudhary on migration patterns (personal communication).

2

TRADE LIBERALIZATION AND SHRIMP EXPORTS

INCREASING OPENNESS OF THE INDIAN ECONOMY IN THE NINETIES: THE MACRO PICTURE

India embarked upon a programme of liberalization in the early 1990s. The reforms involved opening up the economy, reducing the public sector's role, and liberalizing and strengthening the financial sector. The components of this programme were:

a) cutting down fiscal deficit and the rate of growth of money supply so as to keep inflation and balance of payments under control,
b) domestic liberalization consisting of relaxing restrictions on production, investment, prices, and increasing the role of market in guiding resource allocation, and
c) external sector liberalization or relaxing restrictions on international flow of goods, services, technology, and capital.

The aim of the new economic policy was to encourage an increased role of the market in attaining higher rates of growth of the economy. Encouragement was given to expansion of export-oriented units. Licensing for domestic manufacture was abolished for all but a few industries. The private sector was permitted to enter into areas hitherto reserved for the public sector and the Indian rupee was devalued significantly. Some of these policy directions are briefly recapitulated below.

EXCHANGE RATE MOVEMENTS FROM 1991 TO 2003

The single most important factor influencing exports was the exchange rate adjustment, which was a first step towards integrating with the global economy. A two-step downward adjustment of 18–19 per cent in the exchange rate of the rupee was made on July 1 and 3, 1991. In the 1992–3 Budget, Liberalized Exchange Rate Management System (LERM) was introduced along with the dual exchange rate system implying partial convertibility of the rupee. The 1993–4 Budget introduced full convertibility of the rupee on trade account and switched over to a unified exchange rate system. India achieved full convertibility on current account on August 19, 1994.

From the viewpoint of examining the impact of external transactions on the exchange rate stability, the period starting from March 1993 (when the exchange rate became market-determined) can be divided into four sub-periods. During 1993–4, 1994–5 and the first-half of 1995–6, the Indian economy experienced surges in capital inflows with robust export growth, exerting upward pressure on the exchange rate. The nominal exchange rate of the rupee vis-à-vis the US dollar remained virtually unchanged at around Rs 31.37 per dollar over the period March 1993 to August 1995.

The exchange rate of the rupee depreciated by 9 per cent during the period August 1995 to October 1995 before stabilizing in February 1996. The two main factors responsible for this phenomenon were slowing down of capital inflows in the wake of the Mexican crisis and the rise of the US dollar against other major currencies.

In the period from 1997 till 2000, several factors such as the Asian financial crisis, the Russian crisis during 1997–8, increase in international crude oil prices in 1999–2000, and movements in interest rates in the industrialized countries impacted the foreign exchange market. As a result of all these developments, there was a considerable degree of uncertainty in the foreign exchange market leading to excess demand for the dollar. Quick response by the Reserve Bank in terms of a package of measures and liquidity operations, however, saved the situation.

The period from 2002 onwards marked the appreciating trend of rupee against the dollar. The Indian rupee appreciated by 5 per cent in 2004 over 2003 against US dollar from Rs 48.39 per dollar in 2002–3 to Rs 45.95 in 2003–4. However, the Indian rupee depreciated against other key currencies such as Euro, Yen and Pound sterling. On balance, during this period, exporters' expectations with respect to the direction of change in the economy were optimistic with a reasonably stabilized exchange rate ensured in a well-integrated economy.

Other factors contributing to export optimism simultaneously, India became a founder member of the World Trade Organisation (WTO) by ratifying the Uruguay Round GATT Agreement on January 1, 1995. As a founder member of the WTO, India was under an obligation to strike down all quantitative restrictions on imports and reduce import tariffs so as to 'open up' the economy to world trade. Acting on its commitment to the WTO, the Exim Policy of 2001–2 withdrew quantitative restrictions from all import items. Import tariffs were drastically reduced and more liberal imports of a number of goods whose imports were earlier either totally banned or severely restricted have been allowed. The aim of the New Economic Policy was to generate growth, relying on market forces. Focus was given to expansion of export-oriented production.

Factors Impacting Marine Products Exports from India and West Bengal

Among other commodities, marine products exports from India increased substantially in these years. Shrimp production and export from West Bengal too witnessed a substantial increase in the 1990s.[1] In addition to the triggering effect of the exchange rate, the following factors can be stated to have encouraged and facilitated this trend, both for India, and in particular for the state of West Bengal:

1. New mindset among decision makers: Prevalence of a new mindset with optimism for 'export-led growth' thinking among the decision makers in the state of West Bengal in the wake of liberalization programmes in the country in general. In fact, it was a big change in the policy circle of the Left-led government of West Bengal who had earlier believed in inward looking, import substitution strategy. Shrimp, a premium product in the international market became an ideal choice for the policy makers in the State.
2. Favourable international market environment: Around the same time in the early 1990s, the bastion of shrimp export—Thailand and Vietnam got a severe jolt in their supply ability caused by

the breakout of a disease destroying shrimp production. This was caused by intensive farming using excessive chemical fertilizers and pesticides. This provided a big opportunity to India to fill this supply-demand gap and West Bengal Tiger Shrimp was a superb replacement.
3. Increased Private Investment in shrimp processing units in and around Kolkata: An increase in this investment was a consequence of the new opportunities perceived by entrepreneurs as a consequence of the changed domestic and international situation. A contributing factor could have been that some exporters (for instance, tea) were looking for avenues for diversification due to the difficulties faced in international markets for those commodities. They saw an opportunity in shrimp processing and export.
4. Initiatives undertaken by several agencies: Along with these factors, other local government agencies especially the Department of Fisheries of the state initiated favourable policies, which helped the production and export of the shrimp. The state government set up the Brackish Water Fish Farmer's Development Agency (BFFDA). The area covered by it includes some part of Midnapur district along with N24P and S24P districts. During 2000–1, around 3788.94 hectares of area has been developed by the BFFDA. This produces 41 thousand tonnes of shrimp in the state annually (Government of West Bengal 2001–2).

On the organizational side, a restructuring of government departments also contributed to the increased focus on aquaculture. The Directorate of Fisheries was put under the jurisdiction of the Department of Fisheries, Aquaculture, Aquatic Resources, and Fishing Harvest (DFAARFH). This restructuring gave an added thrust to shrimp farming exclusively for export purposes. Some of the programmes and policies undertaken to facilitate and accelerate shrimp farming in the state are listed below:

1. Distribution of fishing nets and boats,
2. Extension and demonstration,
3. Fishermen's group and personal accident insurance scheme,
4. Savings-cum-relief schemes,

Table 2.1: Initiatives Taken by the DFAARFH

S.No.	Name of Scheme	Sector	District	Expenditure Target (Rs)	Physical Target
1.	Brackish water fish farming: project to be implemented through BFGA	Centrally-sponsored scheme	N24P S24P	1,350,000 1,350,000	200ha (N24P) 125ha (S24P)
2.	Fishing nets and requisites: subsidy for marine sector	State plan	N24P S24P	700,000 1,900,000	N24P S24P
3.	Infrastructure: in marine and fishing villages	State plan	N24P S24P	2,000,000 5,000,000	Village roads, tube wells and community halls.
4.	Fishing by-products: diversified production	State plan	S24P	1,300,000	Operation of marine *khutis*; tube wells and welfare measures
5.	Minikit: distribution Social Fishery: development	State plan	N24P S24P	389,000 388,000	Distribution of fishery input and development of social fishery
6.	Netmaking: training of fisherwomen	State plan	N24P S24P	60,000 60,000	Involvement of fisherwomen in fishing activities
7.	Minor fishing harbours Fishing landing centres	Centrally-sponsored scheme	S24P	10,000,000	Fish landing centre and minor fishing harbours
8.	Model village	Centrally-sponsored scheme	N24P S24P	870,000 4,300,000	Construction of model village

(Contd.)

Table 2.1 (Contd.)

S.No.	Name of Scheme	Sector	District	Expenditure Target (Rs)	Physical Target
9.	Common facilities centre for processing and preservation: implementation schemes for marine products	NDCD and Minstry of Food Processing Industries (GOI)	Chakgaria In S24P	168,000,000	Processing and preservation of marine products
10.	Savings-cum-Relief: for marine fishermen	Centrally-sponsored scheme	S24P	900,000	Relief to marine fishermen during lean period (March–June)
11.	Aquaculture development	Centrally-sponsored scheme	N24P S24P	3,000,000 3,300,000	Development of aquaculture through bank finance for increasing fish production
12.	Fishermen Group Personal Accident Insurance: through FISHCOPPED in master Policy	Centrally-sponsored scheme	N24P S24P	126,000 150,000 (50% premium on cost)	Accident insurance coverage for death and permanent disablement
13.	Tribal people: economic upliftment through pisciculture and housing	Centrally-sponsored scheme	S24P	400,000	Housing for tribal people

Source: Directorate of fisheries, Government of West Bengal.

5. Training,
6. Provision of housing,
7. Brick pavement links and approach roads,
8. Tank development, and
9. Communication and infrastructure development.

The steps undertaken by the Department are summarized in Table 2.1. Some of these are directed primarily at encouraging shrimp farming and export.

These drivers of change at the international, national, and regional levels resulted in a consolidation of the position of Indian shrimp exports and production in the world economy.

INDIA AND THE WORLD SHRIMP ECONOMY

In India, shrimp aquaculture started in the late 1980, and it became a significant activity in the early 1990s. There have been ups and downs in this activity during this period as expected. The boom period of commercial-scale shrimp culture in India started in 1990 and the bust came in 1995–6, with the outbreak of viral disease. The fact that most of the coastal states in India were new to commercial-scale shrimp farming, the general ignorance of good farming practices, and the lack of suitable extension services, led to a host of problems. Commercial shrimp farming also developed on account of the government's policy to promote shrimp culture in order to provide employment opportunities to the coastal rural population and to earn valuable foreign exchange. After the liberalization of the Indian economy in 1991, individual entrepreneurs were also encouraged to take up shrimp farming with both financial and technical support. As a result, India was among the top shrimp producing countries of the world (Table 2.2).

China is the largest producer of shrimp in the world at 1,241,900 tons in 2000, comprising nearly 30 per cent of the world's production. India is the second largest producer of shrimp in the world with 4,05,700 M tons of shrimp production in 2000. The share of India in the world's shrimp production in 1991 was around 10 per cent, which remained the same in 2000 as well, though production grew at an average rate of 4 per cent over this period. The share of China

Table 2.2: Yearly Shrimp Production by Major Producing Countries 1991–2000

(in '000 M. Tons)

Country	1991	1992	1993	1994	1995	1996	1997	1998	1999	2000
China	564.1	574.1	488.7	603.4	665.6	751.8	829.6	970.9	1,222.7	1,241.9
India	300.5	290.4	363.0	446.6	406.1	415.6	366.6	413.1	423.3	405.7
Thailand	289.9	300.6	343.1	385.0	389.3	370.8	350.8	345.4	370.9	398.5
Indonesia	296.8	312.1	300.7	317.1	334.7	343.3	382.2	345.5	384.5	398.4
USA	148.5	156.5	137.9	130.2	140.2	145.0	132.9	128.0	140.1	153.0
Vietnam	81.3	86.2	94.6	111.7	138.1	135.9	147.7	148.4	148.9	151.1
Canada	44.7	43.1	47.4	53.2	63.1	65.7	82.1	113.1	120.0	130.6
Malaysia	104.7	129.4	109.8	106.4	99.6	108.0	101.0	57.1	102.7	111.9
Mexico	70.6	66.2	79.8	77.3	85.9	78.9	88.5	90.3	95.6	95.1
Greenland	73.1	81.9	76.5	79.8	81.9	72.0	63.9	69.6	79.2	81.5
Philippines	84.9	118.8	130.1	126.6	127.5	113.2	74.5	72.3	73.1	79.4
Norway	49.0	49.1	49.0	38.2	39.3	41.5	42.0	57.1	64.2	66.2
Bangladesh	19.6	21.0	28.5	28.8	34.0	49.3	56.5	66.1	81.1	58.2
Brazil	42.3	44.0	38.4	38.5	43.0	38.9	44.1	42.8	47.7	56.6

(Contd.)

Table 2.2 (Contd.)

Country	1991	1992	1993	1994	1995	1996	1997	1998	1999	2000
Ecuador	118.8	127.0	97.5	98.7	112.1	112.9	137.2	147.4	121.0	51.4
Korea Rep.	55.8	67.1	68.0	58.1	42.5	40.9	41.1	47.6	44.7	37.2
Others	532.7	529.3	542.0	551.7	594.5	622.9	633.7	647.4	599.2	651.7
Total	2,877.3	2,996.8	2,995.0	3,251.3	3,397.4	3,506.6	3,574.4	3,762.1	4,118.9	4,168.4

Source: www.foodmarketexchange.com

Note: data include all types of shrimp, namely farm-raised shrimp and wild shrimp.

in the total shrimp production increased from 20 per cent to 30 per cent in the same period.

In India, shrimp production is highest in Andhra Pradesh followed by West Bengal (Table 2.3). West Bengal with a coastline of 158 kms and a continental shelf of 17,000 sq km produces about 22 per cent of the total production in the country. It is endowed with the highest potential resources of brackish water aquaculture (27 per cent of the country's potential) among all the maritime states. The State's share of the saline soil is about 0.08 million hectare (mha) out of 2.10 mha in the country. The total potential culturable area is estimated at about 0.21 mha and presently 0.048 mha of water area has been brought under brackish water farming (Government of India 2002).

Table 2.3: State-wise Shrimp Production in India

(in M. Tons)

States	1998–9	Share	1999–2000	Share	2000–1	Share
Andhra Pradesh	44,856	54%	46,270	54%	53,100	55%
West Bengal	18,326	22%	21,780	25%	21,079	22%
India	82,634		86,000		97,096	

Source: www.Indiastat.com

The brackish water fisheries development is high in West Bengal particularly because of the extensive saline soil-water resource, human resource, favourable agro-climatic conditions, productive estuarine ecosystems including the Sundarbans, and also abundance of prawns and other brackish water finfish varieties.

MARINE PRODUCTS AND SHRIMP EXPORTS FROM INDIA

Marine products exports are the largest component of India's agricultural and food exports, accounting for approximately 17 per cent of the total in 2003–4. The share of marine products exports in agricultural exports varied from 15 to 20 per cent from 1990–1 to 2003–4. The growth in exports of fish and fishery products outpaced the overall exports of agricultural and food products from 1990–1 to 2003–4. Over this same period, the contribution of marine products

Table 2.4: Share of Marine Products Exports in Agricultural Exports and Total Exports of India

Year	Agricultural Exports Rs Crore	Agricultural Exports US$ Million	Total Exports Rs Crore	Total Exports US$ Million	Marine Products Exports Rs Crore	Marine Products Exports US$ Million	Marine Products Export as a per cent Agricultural Exports	Marine Products Export as a per cent Total Exports
1970–1	487	644	1,535	2,031	3	4	0.62	0.20
1980–1	2,057	2,601	6,711	8,486	217	274	10.55	3.23
1990–1	6,317	3,521	32,553	18,143	960	535	15.20	2.95
1995–6	21,138	6,320	106,353	31,797	3,381	1,011	15.99	3.18
1996–7	24,239	6,828	118,817	33,470	4,008	1,129	16.54	3.37
1997–8	25,419	6,840	130,101	35,006	4,487	1,207	17.65	3.45
1998–9	26,104	6,205	139,752	33,218	4,369	1,038	16.74	3.13
1999–2000	25,016	5,773	159,561	36,822	5,125	1,183	20.49	3.21
2000–1	28,582	6,256	203,571	44,560	6,367	1,394	22.28	3.13
2001–2	29,312	6,146	209,018	43,827	5,897	1,236	20.12	2.82
2002–3	33,691	6,962	255,137	52,719	6,928	1,432	20.56	2.72
2003–4	36,247	7,888	293,367	63,843	6,106	1,329	16.85	2.08

Source: Economic Survey, Various issues.

to total merchandise exports remained constant at approximately 3 per cent.

Even more rapid expansion occurred in the 1990s, and exports of marine products increased more than three times to over 4,12,017 MT from 1,39,419 MT during 1990–1 to 2003–4. This was much more than the average rate of growth of 10 per cent per annum seen from 1960 to 1990. In value terms, growth in exports was equally dramatic from around US $500 million in 1990–1 to US $1,330 million in 2003–4.

Source: *Marine Products Export Review*, MPEDA (various issues).

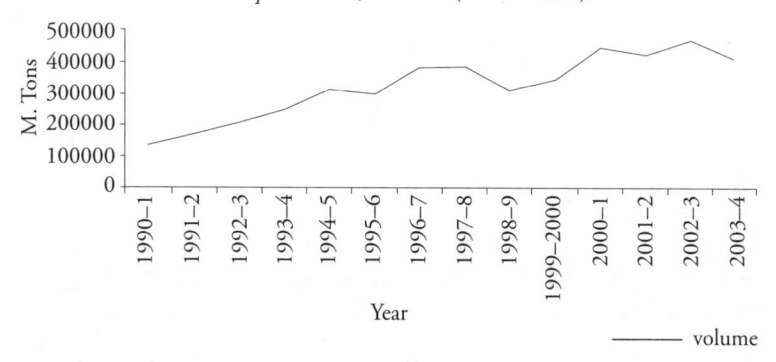

Source: *Marine Products Export Review*, MPEDA (various issues).

Figure 2.1: Export Growth of Indian Marine Products (Volume)

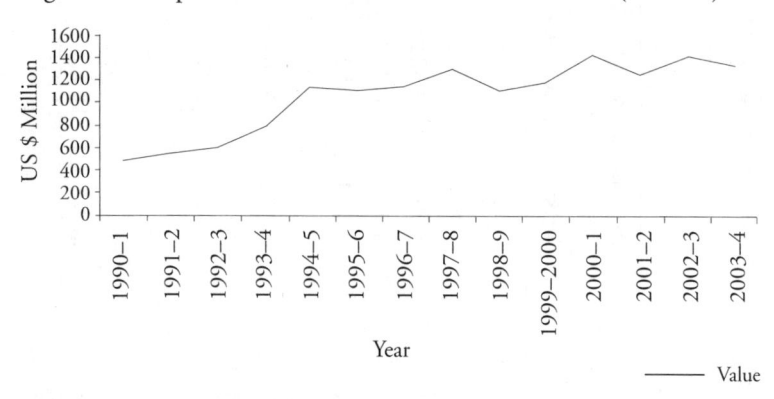

Source: *Marine Products Export Review*, MPEDA (various issues).

Figure 2.2: Export Growth of Indian Marine Products (Value)

1991–2

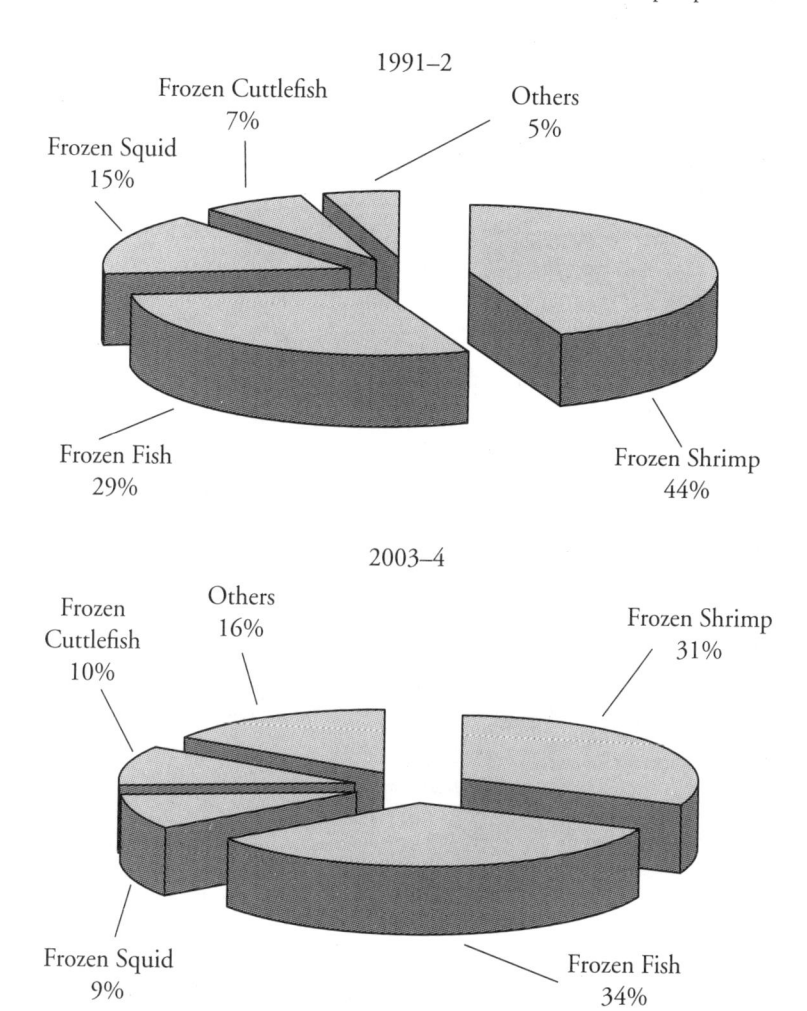

- Frozen Cuttlefish 7%
- Frozen Squid 15%
- Others 5%
- Frozen Fish 29%
- Frozen Shrimp 44%

2003–4

- Frozen Cuttlefish 10%
- Others 16%
- Frozen Shrimp 31%
- Frozen Squid 9%
- Frozen Fish 34%

Source: *Marine Products Export Review*, MPEDA (various issues).

Figure 2.3: Composition of Marine Products Exports from India (Volume)

Figures 2.3 and 2.4 give details of the composition of marine products exports from India from 1991–2 to 2003–4 in terms of volume and value both. The figures show that, frozen shrimp is the major product in the Indian exports of marine products. Over the

1991–2

- Frozen Cuttlefish 4%
- Frozen Squid 8%
- Others 6%
- Frozen Fish 10%
- Frozen Shrimp 71%

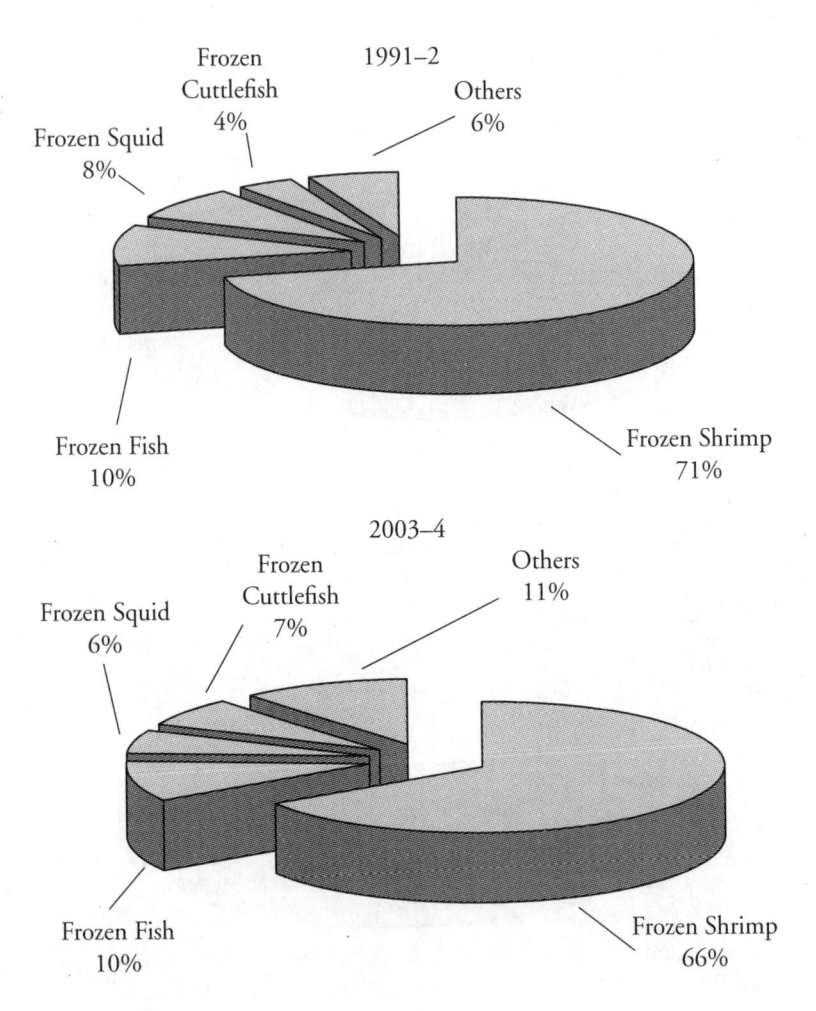

2003–4

- Frozen Cuttlefish 7%
- Frozen Squid 6%
- Others 11%
- Frozen Fish 10%
- Frozen Shrimp 66%

Source: *Marine Products Export Review*, MPEDA (various issues).

Figure 2.4: Composition of Marine Products Exports from India (Value)

period 1991–2 to 2003–4, the value of shrimp exports increased by approximately 150 per cent from US $395.98 million to US $985 million. Throughout this period, although the volume of frozen shrimp exports as a proportion of total fish and fishery product exports has declined from 44 per cent to 31 per cent, by value its share has increased from 66 per cent to 71 per cent. Other significant exports include frozen fish, whose share increased from 29 per cent in 1991–2 to 34 per cent in 2003–4 in terms of volume. By value, the share remained static around 10 per cent over the same period. The share of frozen squid declined both in terms of volume and value but the decline was more by volume from 15 per cent to 9 per cent as against 8 to 6 per cent by value from 1991–2 to 2003–4. On the other hand, the share of frozen cuttlefish increased from 7 per cent to 10 per cent and from 4 per cent to 7 per cent by value during the same period.

European Union (EU), Japan, and the United States, collectively account for approximately 71 per cent by value of the total marine products exports from India. Of this USA alone accounted for 28 per cent followed by EU (24 per cent) and Japan (19 per cent) (see figure 2.5). Through the 1990s, however, there were significant changes in the destination of Indian exports. Most notably, China and (to a lesser extent) other parts of south-east Asia have emerged as important markets that accounted for close to 19 per cent of Indian exports in 2003–4. Furthermore, the share of the United States as an important market has increased from 11 per cent by value in 1992–3 to 28 per cent in 2003–4. Exports of frozen shrimp to the United States also increased from 12 per cent in 1992–93 to 37 per cent in 2003–04.

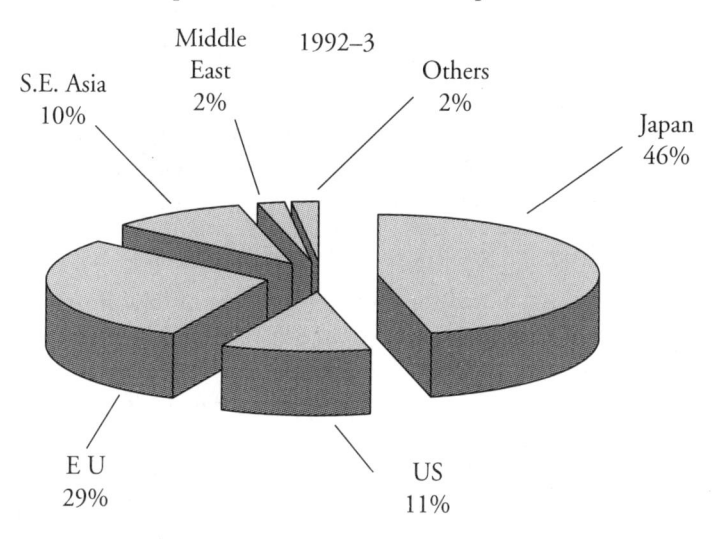

1992–3

S.E. Asia 10%
Middle East 2%
Others 2%
Japan 46%
EU 29%
US 11%

2003–4

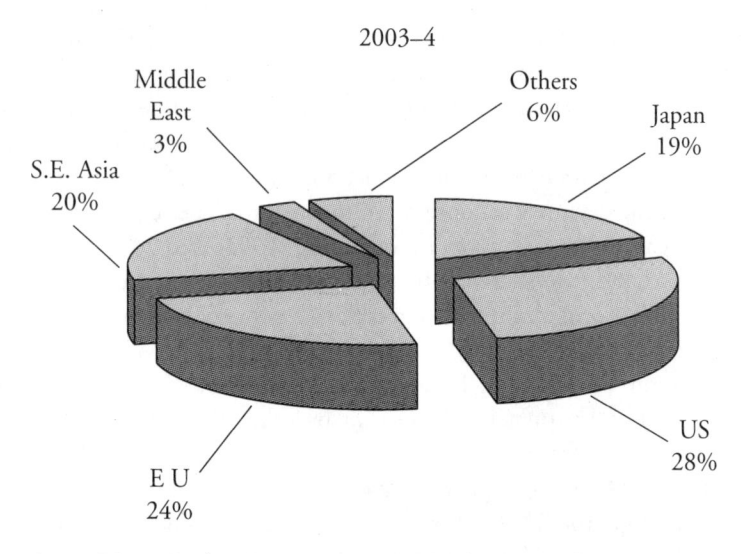

- Middle East 3%
- S.E. Asia 20%
- Others 6%
- Japan 19%
- US 28%
- EU 24%

Source: *Marine Products Export Review*, MPEDA (various issues).

Figure 2.5: Destinations of Marine Products Exports from India (Value)

During the same period, total marine products exports to Japan declined from 46–19 per cent while frozen shrimp exports declined from 61–23 per cent. Although exports of marine products to the EU experienced a decline from 29 per cent in 1992–93 to 24 per cent in 2003–04, frozen shrimp exports declined marginally from 22 per cent in 1992–93 to 21 per cent in 2003–04 (See tables 2A.1a and 2A.1b in Appendix 2A). Over the same period, exports to Japan declined from 46–19 per cent. Exports to the EU have also declined from 29 per cent in 1992–3 to 24 per cent in 2003–4.

Frozen Shrimp Exports from West Bengal

West Bengal accounted for about 11 per cent of total frozen shrimp exports from India in terms of volume in 2003–4. The share increased from 9.3 per cent in 1991–2 to approximately 11 per cent in 2003–4. However, the percentage share by value remained the same, that is, around 13 per cent for the same period. In absolute terms, frozen shrimp exports were Rs 126.33 crores in 1991–2 which increased to Rs 508 crores in 2003–4, an increase of 300 per cent over 12 years.

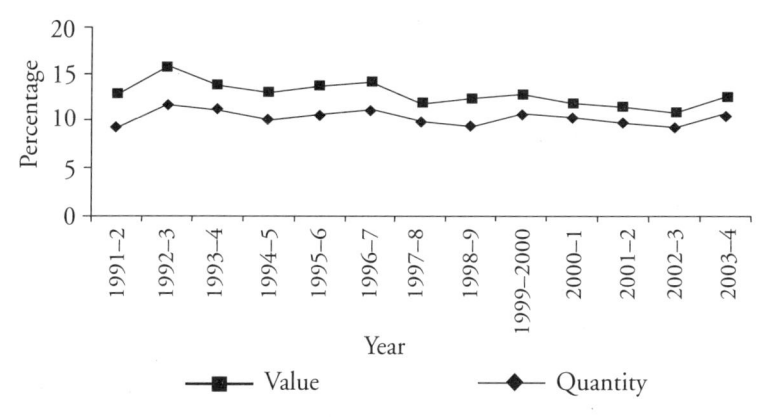

Source: Marine Products Export Review, MPEDA (various issues).

Figure 2.6: Percentage Distribution of West Bengal in the Total Frozen Shrimp Exports from India

The major export destinations in the world for West Bengal's frozen shrimps are Japan, USA, and the EU. These three destinations together account for about 90 per cent of the total exports from West Bengal. Japan was the major export destination of frozen shrimps until 2003 followed by the USA and the EU, but Japan's share declined in later years. In 1995–96, Japan's share was 68 per cent (in terms of value), which declined to 50 per cent in 1999–2000 and further to 38 per cent in 2003–04. On the other hand, the share of USA in the total frozen shrimp exports increased from 7 per cent in 1995–96 to 21 per cent in 1999–00 and finally to 40 per cent in 2003–04. The share of EU increased from 18 per cent in 1995–96 to 22 per cent in 1999–2000 but later declined to 12 per cent in 2003–04 (See tables 2A.2a and 2A.2b in Appendix 2A).

Simultaneously, however, the share of total marine products export from West Bengal in the total marine products exports from India has declined marginally from 10 per cent in 1991–2 to approximately around 9 per cent in 2003–4. The same trend is observed in terms of volume of marine products exports from West Bengal. The decline is from 5 per cent in 1991–2 to 4 per cent in 2003–4.

Together, the above-mentioned trends indicate that frozen shrimp has become a more important component of marine exports from

West Bengal in the decade under consideration. Both volume and value from the state have increased during this period but not as fast as from some other parts of India. Figure 2.6 depicts the state's share of frozen shrimp export in India's export of frozen shrimp.

In terms of volume, the share grew marginally from 9.36 per cent in 1991–2 to 10.73 per cent in 2003–4 with the highest of 11.72 per cent in 1992–3. The share in value terms shows a marginal decline during the same period. The same declined from 12.90 per cent in 1991–2 to 12.66 per cent in 2003–4 with the highest of 15.65 in 1992–3.

CONCLUDING REMARKS

In this chapter we found that India's and West Bengal's major marine products exports destinations are Japan, USA, and the EU. These three destinations constitute the bulk of the developed world. Although India is among the leading producers of shrimp in the world, exports of shrimp pose challenges like, ensuring quality and meeting food safety standards. Thus, it is important to understand international food safety standards and regulations. In the next chapter we analyse the major determinants of shrimp exports from West Bengal.

NOTES

1. See the Note on shrimp export in Appendix 2.

Appendix 2A

FROZEN SHRIMP EXPORTS FROM WEST BENGAL, INDIA, AND WORLD

Table 2A.1a: Major Markets for Indian Frozen Shrimp

Country		1992–93	%	1995–96	%
Japan	Q	34,258.00		41,955.00	
	V	718.31	61	1,426.04	61
EU	Q	20,757.00		29,397.00	
(Western Europe)	V	257.24	22	495.28	21
USA	Q	14,045.00		16,556.00	
	V	145.17	12	290.12	12
S.E. Asia	Q	2,625.00		4,262.00	
Including China	V	19.51	2	75.20	3
and Hong Kong					
Others	Q	2,366.00		3,566.00	
	V	36.60	3	70.26	3
Total	Q	74,051.00		95,724.00	
	V	1,176.83		2,356.81	

Source: *Review on Export of Indian Marine Products*. Various issues, MPEDA.
Note: Q: Quantity in M.Tons
V: Value in Rs Crore

Table 2A.1b: Major Markets for Indian Frozen Shrimp

Country		2003–04	%
Japan	Q	31,578.00	
	V	1,048.00	23
EU	Q	40,389.00	
	V	959.00	21
USA	Q	43,911.00	
	V	1,662.00	37
S.E. Asia	Q	9,327.00	
	V	291.00	7
China & Hong Kong	Q	5,299.00	
	V	96.00	2
Others	Q	15,440.00	
	V	446.00	10
Total	Q	145,945.00	
	V	4,501.01	

Source: *Review on Export of Indian Marine Products*. Various issues. MPEDA.
Note: Q: Quantity in M.Tons
V: Value in Rs Crore
China and Hong Kong reported separately from 1997–8. *Marine Products Export Review*, MPEDA

Table 2A.2a: Export of Frozen Shrimp from West Bengal

		1995–96	%	1999–00	%
JAPAN	Q	5,299.00		5,095.00	
	V	222.14	68	232.60	50
	$	70.52		54.00	
USA	Q	1,192.00		2,542.00	
	V	22.94	7	100.41	21
	$	7.28		23.39	
EU	Q	2,994.00		3,288.00	
	V	59.55	18	106.03	22
	$	18.90		24.63	
Others	Q	749.00		800.00	
	V	21.37	7	31.96	7
	$	6.95		7.35	
Total	Q	10,234.00		11,725.00	
	V	326.00		471.00	
	$	103.65		109.37	

Source: *Marine Products Export Review*, MPEDA.
Note: Q: Quantity in Tons
V: Value in Rs Crore
$: US Dollar in Millions

Table 2A.2b: Export of Frozen Shrimp from West Bengal

		2003-04	%
Japan	Q	5,108.00	
	V	193.63	38
	$	42.06	
USA	Q	4,821.00	
	V	204.62	40
	$	44.38	
EU	Q	2,390.00	
	V	57.25	12
	$	12.47	
Others	Q	1612.00	
	V	52.50	10
	$	11.40	
Total	Q	13,931.00	
	V	508.00	
	$	110.31	

Source: *Marine Products Export Review* (various issues). MPEDA.
Notes: Q: Quantity in Tons
V: Value in Rs Crore
$: US Dollar in Millions

Table 2A.3: Frozen Shrimp Exports from West Bengal, India, and World

Year		West Bengal	India	World	West Bengal exports as per cent in value terms	
					Indian exports	World exports
1991	Q	7,131.00	83,409.00	898,979.00		
	V	102.52	460.69	6242.65	22.25	1.64
1992	Q	8,682.00	78,409.00	927,340.00		
	V	102.99	454.12	6381.99	22.68	1.61
1993	Q	9,770.00	96,130.00	950,663.00		
	V	103.00	594.85	6822.07	17.32	1.51
1994	Q	10,369.00	110,459.00	1,037,094.00		
	V	104.00	802.05	8092.41	12.97	1.29
1995	Q	10,234.00	9,8456.00	953,542.00		
	V	103.65	682.04	8230.98	15.20	1.26
1996	Q	1,1814.00	110,681.00	977,094.00		
	V	107.48	721.01	7681.19	14.91	1.40
1997	Q	1,0134.00	110,605.00	1,012,637.00		
	V	101.75	796.40	8069.86	12.78	1.26
1998	Q	9,738.00	118,073.00	1,116,815.00		
	V	99.57	751.56	7841.08	13.25	1.27
1999	Q	11,725.00	124,515.00	1,046,626.00		
	V	109.37	771.51	7325.71	14.18	1.49
2000	Q	11,626.00	128,198.00	1,112,079.00		
	V	117.25	896.86	8257.84	13.07	1.42
2001	Q	12,526.00	138,836.00	1,231,873.00		
	V	99.67	798.98	8095.60	12.47	1.23
2002	Q	12,626.00	166,258.00	1,280,550.00		
	V	103.95	889.08	7559.30	11.69	1.38
2003	Q	13,931.00	158,768.00	1,409,068.00		
	V	110.31	825.47	8256.41	13.36	1.34

Source: MPEDA various issues and *www.fishstat.com*.
Notes: Q: Quantity in M. Tons
V: Value US Dollar in Million

3

DETERMINANTS OF SHRIMP EXPORT FROM INDIA AND WEST BENGAL
ANALYSIS AND SOME ECONOMETRIC EXPLORATIONS

INTRODUCTION

The last chapter maintained that the increasing openness of the Indian economy, in particular the exchange rate adjustments, led to increased export of frozen shrimp from India in general and West Bengal in particular. However, as we shall see below, there were ups and downs during this period too and changes in country destinations of these exports as well. What are the factors responsible for this? And how should export-oriented processing units deal with them?

The factors responsible for these changes give us pointers towards future prospects of shrimp exports from the State. Trade liberalization also opens up the economy to the more stringent food, health, and environmental standards of importing countries. These standards vary both from country to country and from commodity to commodity. They may be particularly important for commodities such as frozen shrimp, particularly when imported to developed economies such as the US, EU, and Japan. One strand in the recent literature on export-determining factors[1] maintains that such standards—

- in the short run, act as non-tariff barriers, countering the positive effect of the price variables and the removal of quantitative restrictions, and

- in the long run, they ensure the introduction of universally accepted environmental standards in production and processing in exporting countries and thereby protect their environment and ecosystems.

This chapter intends to examine the impact of such food, health and environmental standards, otherwise referred to as 'non-tariff barriers' on the magnitude and direction of the export of shrimp from West Bengal. Such an investigation is important as it throws light on trade and environment-related issues. To us, its significance also derives from the fact that it helps to determine the permanence or otherwise of the driving factor behind seemingly large changes both in the state's and the region's economy and ecology.

Food, environment, and health standards vary from country to country and also over time. In this chapter, we construct a country-specific index of these. Variations in the index over time are captured through an analysis of the nature and stringency of these restrictions over the time period being studied. The hypothesis postulated is that *exports of frozen shrimp from West Bengal depend on some combination of the following factors: exchange rate, relative prices in destination markets, domestic production of shrimp, real income of importing country, and the environment-related non-tariff measures imposed by importing countries (as approximated by this index)*. The section below recapitulates the magnitude and destination of countries for export of shrimp from West Bengal, followed by the section which describes the methodology and data sources and the next presents the results.[2]

MAGNITUDE AND DESTINATIONS OF EXPORTS OF SHRIMP FROM WEST BENGAL

Frozen shrimp dominates the marine products exports from West Bengal. Its exports grew at an average annual rate of 4 per cent from 1991–92 to 2003–04. In terms of value, exports grew at an average annual rate of 10 per cent over the same period. However, the share of frozen shrimp in total marine products exports from West Bengal declined from 96 per cent in 1991–92 to approximately 85 per cent in 2003–04. In terms of volume, the decline in the share was from

90 per cent in 1991–92 to 75 per cent in 2003–04 resulting in an increase in the share of frozen fish, from around 10 to 12 per cent by volume and, from around 3 to 4 per cent by value over the same period. The share of other items in the total marine products exports grew significantly from 0.04 per cent in 1991–92 to 13 per cent in 2003–04 by volume and from 0.043 per cent in 1991–92 to around 11 per cent in 2003–04 by value (see figure 3.1).

Source: www.mapsofindia.org

Map 3.1: Location of West Bengal in India

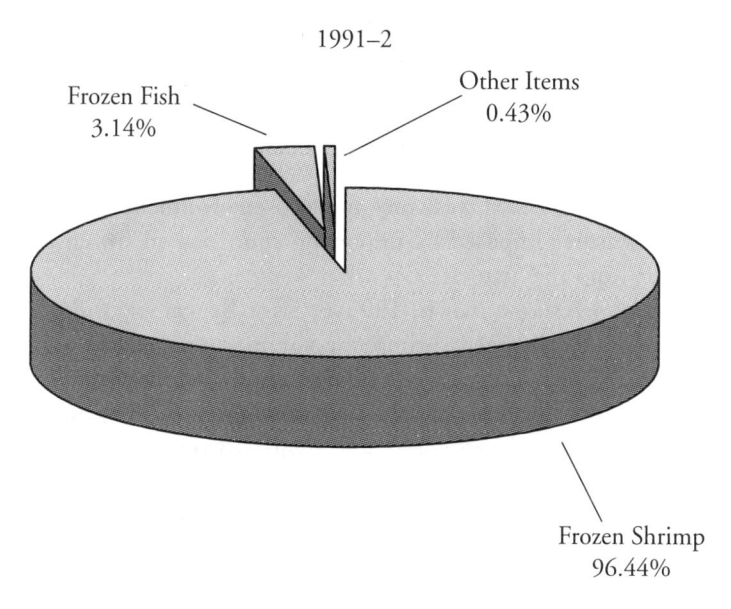

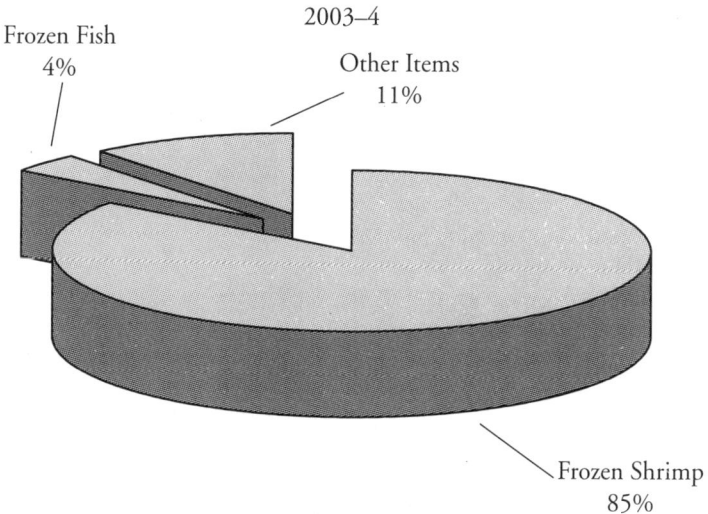

1991–2

- Frozen Fish 3.14%
- Other Items 0.43%
- Frozen Shrimp 96.44%

2003–4

- Frozen Fish 4%
- Other Items 11%
- Frozen Shrimp 85%

Source: *Marine Products Exports Review*, MPEDA (various issues).

Figure 3.1: Composition of Marine Products Exports from West Bengal (Value)

48 Biodiversity, Land-use Change, and Human Well-being

Given this dominance of frozen shrimp in the total marine products exports, it would be interesting to investigate the factors that impact the shrimp exports from West Bengal. In order to do so, an export-demand function has been built up to analyse the impact of prices (relative), shrimp production from West Bengal, exchange rate (Rs/$), real income, and environment-related non-tariff measures imposed by importing countries on frozen shrimp exports from the State.

Figure 3.2 shows that in 1995–6, West Bengal exported 52 per cent of its total frozen shrimps by volume to Japan. EU was the second largest destination with 29 per cent share followed by the US (12 per cent). In terms of value, Japan's share was 68 per cent, EU again the second largest destination with 18 per cent followed by the US (7 per cent). However, in 2003–4, US emerged as the largest export destination with 40 per cent share by value in total frozen shrimp exports. Japan's share declined to 38 per cent followed by EU (11 per cent). But in terms of volume of shrimp exported, Japan was leading with 37 per cent share followed by the US (35 per cent) and EU (17 per cent).

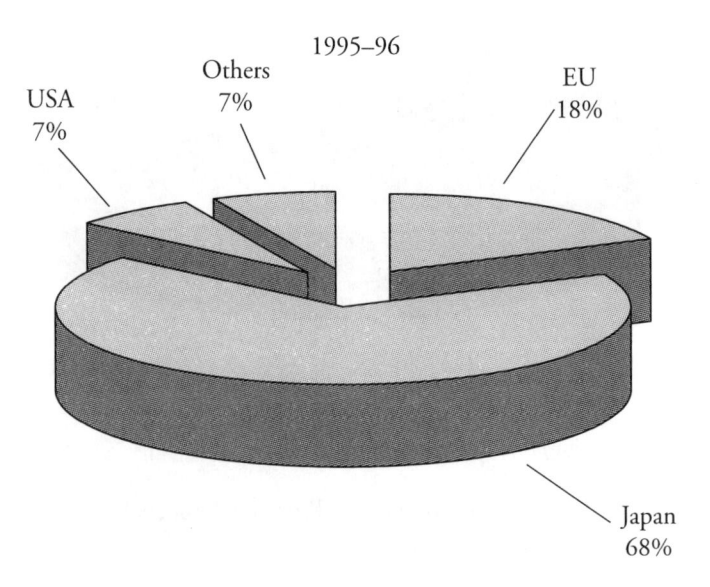

1995–96

USA 7%
Others 7%
EU 18%
Japan 68%

2003–4

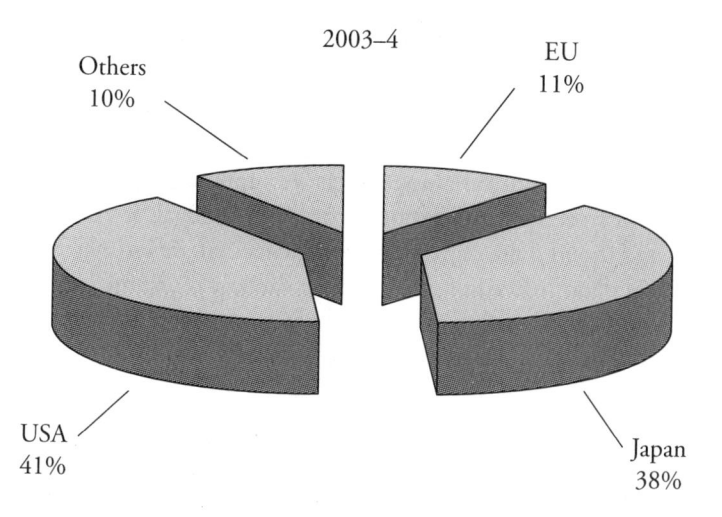

- Others 10%
- EU 11%
- USA 41%
- Japan 38%

Source: Marine Products Export Reivew, MPEDA (various issues).

Figure 3.2: Major Destinations of Frozen Shrimp Exports (Value) from West Bengal

EU is further broken down into UK, Netherlands, and Belgium, based on their share in the total frozen shrimp exports to EU. In 2003–4, these countries together comprised 70 per cent of total frozen shrimp exports by volume and 73 per cent by value to the European Union. The average share of UK in the total frozen shrimp exports to EU from 1995–6 to 2003–4 was 40 per cent and 46 per cent in terms of volume and value—making UK the largest export destination within EU. Belgium is the second largest destination with an average export share of 33 per cent by volume and 32 per cent by value. Netherlands' average share over this period is 11 per cent in terms of volume and 6 per cent by value.

MODEL SPECIFICATION AND METHODOLOGY

The Model

We postulate that export of frozen shrimp from West Bengal depends on production, exchange rate, real income of the importing country, importing country-specific relative prices, and NTMs.

A Random Effects Model has been used in this analysis with the following functional form:

$$Y_{it} = \alpha_i + \beta X_{it} + \mu_{it} \quad (1)$$

X is the matrix of variables, which are supposed to influence the volume of frozen shrimp exports from West Bengal, μ_{it} is an error term, and α and β are regression coefficients that are estimated from the panel data. Instead of treating α_i as fixed, it is assumed that it is a random variable with a mean value of (α). And the intercept value for an individual country can also be expressed as:

$$\alpha_i = \alpha + \varepsilon_i \qquad i=1, 2, \ldots, n \quad (2)$$

where ε_i is a random error term with a mean value of zero and variance of σ^2 and captures the individual differences in the intercept values of each country.

Substituting (2) into (1), we obtain

$$Y_{it} = \alpha + \beta X_{it} + \varepsilon_i + \mu_{it} \quad (3)$$

$$Y_{it} = \alpha + \beta X_{it} + \omega_{it} \quad (4)$$

where

$$\omega_{it} = \varepsilon_i + \mu_{it}$$

The composite error term ω_{it} consists of two components, ε_i which is the cross-section, or individual-specific, error component, and μ_{it}, which is the combined time-series and cross-section error component. The model is also called error component model (ECM). Initially, regressions with different combinations of index of non-tariff measures, relative prices, production, world import price of frozen shrimp, and exchange rate as explanatory variables were tried. But it was found that there is very high negative correlation between production and world import price, high positive correlation between NTMs and production, high negative correlation between NTMs and world prices, and finally high positive correlation between NTMs and exchange rate.

We, therefore, decided to drop world import price, exchange rate, and production from our analysis. Since the study looks at the supply of shrimp from West Bengal, the issue of relative production costs of Indian competitors is not studied. It was felt that this was appropriate in view of the focus of the study being inter-country differences between importing countries. Our analysis is based on the following three regressions:

Log (Frozen shrimp exports from West Bengal to all the five countries) = $\alpha + \beta_1$ log (Index of NTM) + β_2 log (Relative price) + β_3 log (Real income of the importing country)

The dependent variable here is the volume of frozen shrimp exported to all the five countries taken in the study. The analysis rests on the assumption that the exports are determined by food safety standards set by the importing countries along with the relative import prices in these countries. Given the nature of the product being exported, that is, frozen shrimps, qualitative restrictions in the form of NTMs (or food safety standards) are expected to have a significant impact on the volume demanded/exported.

Log (Frozen shrimp exports from West Bengal to all the five countries) = $\alpha + \beta_1$ log (Index of NTM) + β_2 log (Unit value of shrimp imports from Indian competitor in each of these five countries) + β_3 log (Real income of the importing country).

In this regression, the relative price variable is broken down into individual prices. With the same dependent variable as in the above regression, it is assumed that volume exported would depend upon the NTMs (in the form of food safety standards) imposed by the importing countries and on the price these countries have to pay for importing frozen shrimp from their major import markets besides India. For instance, in the US, frozen shrimp imports are highest from Thailand. In this case Thailand is the most important competitor of India.

Log (Frozen shrimp exports from West Bengal to all the five countries) = $\alpha + \beta_1$ log (Index of NTM) + β_2 log (Unit value of shrimp imports from India in each of these five countries) + β_3 log (Real income of the importing country)

The final regression assumes that NTMs along with the import price of Indian frozen shrimp in the destination market (US, Japan, UK, Netherlands, and Belgium) are the important determinants explaining the changes in volume exported from West Bengal.

The Database

The pooling of data was considered a better option than individual country-wise regressions. We have five cross-sectional units (USA, Japan, UK, Netherlands, and Belgium) and nine years time-series (annual data for 1995–2003) observations on the four independent variables and the dependent variable. Therefore, we have a total of 45 pooled observations.

The most important advantage of using panel data sets over traditional pure cross-section or pure time-series data is that the number of observations is typically much larger in panel data. This is likely to give more reliable parameter estimates and specify and test more sophisticated models, which incorporate less restrictive behavioural assumptions. Panel data sets may also alleviate the problem of multi-collinearity.

While it is possible to use ordinary multiple regression techniques on panel data, they may not be optimal. The estimates of coefficients derived from the regression may be subject to omitted variable bias—a problem that arises when there is some unknown variable or variables that cannot be controlled for and that affects the dependent variable. With panel data, it is possible to control some types of omitted variables even without observing them, by analysing changes in the dependent variable over time. This controls for omitted variables that differ between cases but are constant over time. It is also possible to use panel data to control for omitted variables that vary over time but are constant between cases.

Description and Construction of Variables

Dependent Variable

LnEX- log (Volume of frozen shrimp exports from West Bengal to all the five countries)

Independent Variables

LNTM—log (Index of non-tariff measures)

LnRP—log (relative prices)

LnCPR—log (Price of shrimp imports from India's major competitor in the US, Japan, UK, Netherlands, and Belgium)

LnINDP—log (Price of shrimp imports from India in the US, Japan, UK, Netherlands, and Belgium)

LnRGDP—log (Real GDP of the importing countries, all in US $)

Data Construction

Dependent Variable

EX—Volume of frozen shrimp exported from West Bengal(tons)[3]

Independent Variables

RP—Relative Price—For each of the five countries taken in our analysis, India's major competitor for frozen shrimp in that particular market is identified and the relative price calculated for these five countries separately. For instance, in the US market, relative price would be determined by taking the unit value ($/Kg) of frozen shrimp imports from India relative to unit value of its major competitor in the US (Thailand). Imports of US frozen shrimp are the highest from Thailand. Similarly, for other markets, major competitors were identified and the relative prices calculated.[4] The term price used in the analysis means the unit value given in $/Kg.

Major Destinations	*India's major competitor in these destinations*
Japan	Indonesia
UK	Bangladesh
Netherlands	Nigeria
Belgium	Bangladesh
US	Thailand

NTM—Non-tariff Measures—A country-specific index of non-tariff measures has been constructed for the US, Japan, and EU separately. These measures affect the processing units and not the shrimp farmers. The index built for EU was applied to UK, Netherlands, and Belgium. This index was built on the basis of information taken from various official websites of importing countries. This primarily involved documenting the food safety standards laid down from time to time by these importing countries. The amendments made in the existing standards, making them more stringent with every new amendment, were also studied to provide inputs on changes over time. A detailed note explaining the various NTMs and the method for generating the index has been explained in Annexure 3A.

CPR ($/Kg)—Competitor's Import price (Unit value)—Price of frozen shrimp imports from major importing countries, besides India, in the US, Japan, UK, Netherlands, and Belgium separately.[5]

INDP ($/Kg)—Indian import price (Unit value)—Price of frozen shrimp import from India in the US, Japan, UK, Netherlands, and Belgium.[6]

RGDP (US $)—GDP at current prices deflated by the implicit price deflator series (base year 2000) given in US $ for all the five countries. Data taken from the International Financial Statistics (IFS), IMF.

MODEL RESULTS AND INTERPRETATION

Model 1: Log (Frozen Shrimp Exports from West Bengal to all the five countries) = $\alpha + \beta_1$ log *(Index of NTM)* + β_2 log *(Relative price)* + β_3 log *(Real income of the importing country)*

Table 3.1: Export of Frozen Shrimp from West Bengal (Random Effects Model)

Variable	Random Effects
Log NTM = Non-tariff measures	–2.02
	(–1.92)
Log RP = Relative Price	3.29*
	(2.4)
Log RGDP= Real income	0.25
	(0.45)
Constant	15.24
Breusch and Pagan Lagrangian Multiplier (LM) Chi²	6.24
(Probability LM Chi²)	(0.021)
Hausman Chi²	0.9
(Probability Hausman Chi²)	(0.825)
Adjusted R²	0.373
N (Number of groups = 5, Observations per group=9)	45

Source: Authors' calculations.
Note: * denotes significance at 5 per cent.

The Breusch Pagan Lagrangian Multiplier (LM) test and the Hausman test are performed in order to guarantee the robustness of empirical results.[7] The result tells us that the classical regression model with a single constant term is inappropriate for our data and favours the Random Effects Model.

The specification of the model is tested by using the Hausman test. The critical value from the Chi^2 table with three degrees of freedom is 7.81, which is larger than the test statistic, 0.9. The hypothesis that the individual effects are uncorrelated with the other regressors in the model cannot be rejected. The test suggests that of the two alternatives considered, the Random Effects Model is the better choice.

The Random Effects Model shows overall good fit with the correct sign for the index of non-tariff measures, which is significant at around 5 per cent (5.5 per cent). The variable relative price, though significant at 5 per cent, has the sign contrary to what is expected. On the other hand, the variable real income of the importing country has the correct sign but is insignificant at 5 per cent. In the initial run, shrimp production from West Bengal was also taken as one of the explanatory variables, but it came out to be insignificant with high positive correlation with another variable NTM and, therefore was dropped from the final regression.

In view of the above results, which indicate that 'relative price' is not a significant explanatory variable, the following specification takes the competitor's price as the independent variable, instead of the relative price. The Breusch Pagan LM test (Tables 3.2a, b) is used in order to choose between a simpler panel-data model (classical regression) without effects and a model with fixed or random effects.[8] We find that the classical regression model with a single constant term is appropriate for our study.

Model 2: Log (Frozen shrimp exports from West Bengal to all the five countries) = α + β_1 *log (Index of NTM)* + β_2 *log (Unit value of shrimp imports from India's competitor in each of these five countries)* + β_3 *log (Real income of the importing country)*

The results of the classical regression model show that both the variables, index on non-tariff measures and competitor's price (CPR)

Table 3.2a: Export of Frozen Shrimp from West Bengal (Classical Regression Model)

Variable	Coefficients
Log NTM = Non-tariff measures	−1.367
	(−1.23)
Log CRP = Competitor's Price	0.971
	(0.45)
Log RGDP= Real income	0.67*
	(2.74)
Constant	6.84
Breusch and Pagan Lagrangian Multiplier (LM) Chi^2	0.22
(Probability LM Chi^2)	(0.642)
Adjusted R^2	0.242
N (Number of groups = 5, Observations per group=9)	45

Source: Authors' calculations.
Note: 1. t statistics is in the parenthesis.
2. * denotes significance at 5 per cent.

are insignificant at 5 per cent. The only significant variable in this regression is the real income of the importing country. Further, all the variables have the expected signs.

Model 3: Log (Frozen shrimp exports from West Bengal to all the five countries) = α + β_1 log (Index of NTM) + β_2 log (Unit value of shrimp imports from India in each of these five countries) + β_3 log (Real income of the importing country)

The results of the Lagrangian Multiplier (LM) test statistic of 0.09, is less than the 95 per cent critical value for Chi^2 with one degree of freedom, 3.84. We conclude that the classical regression model with a single constant term is appropriate for our study.

Based on the LM test, classical regression is used. The sign of the price variable (INDP) is correct and is significant (at 5 per cent level). The other variables, index of non-tariff measures and real income of the importing country, are insignificant with the correct signs.

Table 3.2b: Export of Frozen Shrimp from West Bengal
(Classical Regression Model)

Variable	Coefficients
Log NTM = Non-tariff measures	−0.712
	(−0.74)
Log INDP = Price of shrimp imports from India	3.6*
	(2.42)
Log RGDP= Real income	0.11
	(0.34)
Constant	6.835
Breusch and Pagan Lagrangian Multiplier (LM) Chi^2	0.09
(Probability LM Chi^2)	0.77
Adjusted R^2	0.334
N (Number of groups = 5, Observations per group = 9)	45

Source: Authors' calculations.
Note: 1. t statistics is in the parenthesis.
2. * denotes significance at 5 per cent.

DISCUSSION AND CONCLUDING REMARKS

On the whole, the statistical results lead towards mixed conclusions but examining their nature is of some interest. The results from the first regression show a statistically significant relationship of index of non-tariff measures and relative prices with the frozen shrimp exports. The sign of the coefficient of index of non-tariff measures is negative, as expected, which tells us that as the food safety standards in the importing countries become more stringent, the volume of shrimp exported from West Bengal goes down. The value of −2.06 of the NTM coefficient implies that a one-percentage point increase in the index of non-tariff measures would decrease the exports by 2 per cent approximately. In other words, the associated increased cost of complying with these standards may have led to a fall in the profits of some of the existing units, thereby inducing the firms (especially small-scale units) to reduce their operations or close down completely. Also, some units could have diverted their exports to other markets in order to avoid incurring huge costs in terms of infrastructure required in meeting these standards. Another important explanation could be that inspite of making huge investments, the rejection or detention rates for export consignments might have increased since it takes time for the processing units to understand the exact requirements, during which the products exported might get rejected or detained, and firms upgrade their units accordingly.

The coefficient of the relative price variable has a positive sign implying that volume exported from India increases even when the price of Indian imports increases relative to that of the competitor in the destination markets. The variable is highly significant at 1 per cent. This is quite contrary to what is expected—that is, as the import price in the importing countries increases, the demand for our product should go down, and as a result, the volume exported would also decrease.

Based on the above obtained results, it was thought that separate regressions capturing the effect of change in the prices, competitor's price (India's competitor in the importing market), and the shrimp import price from India in these five importing countries, on our exports should also be studied. The variable relative price was dropped and instead two regressions—one with competitor's price and the other with India's import price in the importing country were run.

In the first regression with competitor's price, both the variables, non-tariff measures and competitor's price, came out to be insignificant at 10 per cent, with correct signs and high negative correlation.

In the second regression, the price of imports from India (INDP) in the destination market is significant but with incorrect sign. The variable, NTM, is insignificant with correct sign. The correlation between NTM and INDP, though negative, is not high.

Also, it is important to note that the introduction of real income of an importing country as an explanatory variable improves the explanatory power of the model uniformly. It turns out to be significant only in conjunction with 'the competitor's price' version of the price variable. It also does not influence the signs of other variables as expected.

However, even as different versions of the price variable turn out to be either insignificant or resulting in counter-intuitive results, the food safety standards, acting as NTMs, do seem to have a negative effect on the volume exported.

The indefinite nature of the results indicates that the underlying relationships defining the export supply or the import demand equations are not captured well by the panel-data approach. This may be due to the underlying relationships being different in different countries. This implies that the inconclusive results are due to inadequate data. An examination of individual country graphs on index of NTMs and the exports value for West Bengal indicates that they move in opposite directions for some years and for some countries (see graphs in Annexure 3.C). Exports fall as a consequence of stringent measures. Then they pick up again as the process of adjustment by exporting units continues. It seems that there is a continuous process of adaptation going on as firms seek to comply by absorbing new knowledge on export requirements and incurring the necessary expenditures in processing, labelling, transport etc. to do so. As long as the international prices are remunerative enough to enable a fair margin of profit after this compliance cost is internalized, the adjustment process will continue as indeed the increasing exports of shrimp indicate.

A non-linear adjustment to increasing stringency of NTMs takes place over time and linear relationships implicit in the regression models considered do not capture them. Further, our in-depth interviews with processing units indicate the following channels of impact:

1. Opening up of world trade through a simultaneous reduction of tariffs helps in the process by keeping the international prices high, and
2. Timely availability of complete information on the nature of import requirements by different countries is an important contributing feature that enables this adjustment process to move along without discontinuities marked by events such as rejection of consignments. Such discontinuities disrupt adjustment processes and result in loss of livelihoods all along the chain from processing units to collectors of prawn seed, and
3. Government policies like financial support to the processing firms to comply with NTMs, export subsidies, and administrative help in dealing with detention and rejection of firms' shipments by importing countries would facilitate the adjustment process.

NOTES

1. It has been argued that countries heavily dependent on exports incur a welfare loss when environmental standards, regulations, and protocols are introduced by importing countries. See Ulph (1999), Markusen (1999),Bhagwati (1963). For industry-specific studies in India, see Mehta and George (2005), Kadekodi and Mishra (2003), and Chopra and Agarwal (1999).
2. Details with respect to construction of the variable series, in particular the index of non-tariff barriers are given in the appendices.
3. Data taken from MPEDA (Cochin) for the period 1995–6 to 2003–4.
4. Data taken from UN Comtrade.
5. Data taken from UN Comtrade.
6. Data taken from UN Comtrade
7. The LM test statistic of 6.24 exceeds 95 per cent critical value of Chi^2 with one degree of freedom (k-1), 3.84.
8. The LM test statistic of 0.22 is less than the critical value for Chi^2 with one degree of freedom, 3.84.

Appendix 3A

CONSTRUCTION OF THE INDEX OF NON-TARIFF MEASURES

3A.1 INTERNATIONAL FOOD SAFETY STANDARDS AND THEIR IMPACT ON INDIAN SEAFOOD INDUSTRY

Inspite of the lifting of quantitative restrictions and tariffs in other countries during the 1990s, gains from market access under WTO negotiations were not significant as (Kumar and Kumar 2003), the lifting of tariff barriers was accompanied by the introduction of technical regulations and standards. Trade barriers to fisheries products are in the form of non-tariff measures including food safety standards, labelling, and certification requirements, the most important being food safety standards.

3A.1.1 Sanitary and Phyto-sanitary Measures (SPS) and Technical Barriers to Trade (TBT)

The Agreement on Sanitary and Phyto-sanitary (SPS) measures is the most important requirement of food safety. The SPS Agreement attempts to address the application of measures associated with the protection of human, animal, and plant health in such a way so as to prevent such measures from being used as unjustified trade barriers. The Agreement stresses that not only should SPS measures be scientifically based but also stresses on the importance of risk assessment in determining the appropriate levels of these measures. Also, there should be complete transparency in the development and implementation of measures and the adoption of international standards. The SPS Agreement is based on the standards and guidelines established by the Codex Alimentarius Commission for food additives, veterinary drug and pesticide residue, and codes and guidelines of hygienic practice.

The SPS Agreement applies only to measures covering food safety, animal and plant life, and human health. Other technical measures outside this area come within the scope of the Agreement on Technical Barriers to Trade (TBT Agreement). The SPS and TBT Agreements are thus complementary and mutually reinforcing. Technical regulations and standards are used extensively for fish trade and constitute barriers to trade. The TBT Agreement ensures that countries do not use technical regulations and standards, such as packaging, marking, and labelling requirements, as barriers to international trade discriminating in favour of domestic producers or goods of different origin.

The issue of sanitary and phyto-sanitary measures has been a major cause for concern for the Indian seafood industry since August 1997, when the

European Union (EU) banned the import of seafood from India. The ban was imposed due to serious deficiencies with regard to infrastructure and hygiene in fishery establishments and potential high risk to public health with regard to production and processing of fisheries products.

3A.1.2 Eco-labelling

Labelling is another non-tariff measure that acts as a significant trade barrier. Complications in labelling requirement increase the cost of exports. Although labelling involves some additional costs for the exporter, the label leaves the final decision on buying to the consumers. This also applies to labelling requirements in process and production methods (PPM) covering environmental issues. Eco-labelling requirements have to be in line to the provisions in the Agreement on Technical Barrier to Trade. The Agreement specifies that such requirements should be transparent and that notice should be given to the exporter prior to shipment, allowing adequate time to meet the requirement.

3A.1.3 Certification Requirement

Various certificates placing emphasis on quality control, fish handling, and processing are required by the importing countries, for example, health certificates, sanitary certificates, certificates on HACCP, ISO 9000, ISO 14000, ISO 14001, and MSC certificates. This certification requirement could act as an obstacle to international trade.

3A.1.4 EU Regulations

EU standards are enforced and regulated at the country level and thus a restriction of exports to the EU under the regulations affects the entire export community. On the other hand, food safety regulations imposed by other countries like Japan and the US are enforced on a unit and so a restriction on imports affects one particular exporter. Moreover, a country has to be licensed to export to the EU, also, each individual exporting company has to apply to the 'competent authority' within the exporting country for permission to do so. Hazard Analysis Critical Control Point (HACCP) is also a part of the EU food safety standard, but the EU standards are higher than the HACCP standards, so those plants approved for export to the EU could also export to the US.

3A.1.5 US Regulations

The US Food and Drug Administration (FDA) enforces the HACCP requirements by examining products at the point of entry and also inspects the importer's place of operation, if deemed necessary. The regulations are imposed so that the purchaser/importer of the products demonstrates to the authorities that the products have been produced in a safe and acceptable manner.

3A.1.6 Japanese Regulations

The Food Sanitation Law and the Quarantine Law govern standards for imports of fish and fishery products into Japan. These laws prohibit *inter alia* the imports for sale of unsanitary foods, foods not conforming to prescribed standards of manufacture, and composition. The consignments are checked for signs of decomposition, level of total basic nitrogen, and for the presence of foreign matter. Besides all the above, they are also checked for the presence of contaminants such as antibiotic residues, mercury, pesticides, etc.

3A.2　INDIAN FOOD SAFETY REGULATIONS

In India, the Bureau of Indian Standards (BIS) has been designated as the WTO-technical Barriers to Trade Enquiry Point, while the Ministry of Commerce is responsible for implementing and administering the WTO agreement on TBT. Indian standards are formulated by the BIS, which endeavours to align Indian standards as far as possible with international standards.

The Codex Alimentarius Commission (CAC) created by Food and Agriculture Organization of the United Nations (FAO) and the World Health Organization (WHO) in 1961/62 helped in developing food standards, guidelines, and related texts such as codes of practice under the Joint FAO/WHO Food Standards Programme. The main purpose of this programme is to protect the health of consumers, ensure fair practices in the food trade, and promote coordination of all food standards work undertaken by international governmental and non-governmental organizations. It has become a global reference point for consumers, food producers and processors, and national food control agencies. Codex India, the National Codex Contact Point (NCCP) for India, is located at the Directorate General of Health Services, Ministry of Health and Family Welfare (MOHFW), Government of India. It coordinates and promotes Codex activities in India in association with the National Codex Committee and facilitates the country's input to the work of Codex through an established consultation process. India became the member of the CAC in 1970.

The Export Inspection Agency (EIA) is in overall charge of EU certifications to be accorded to the processing units. Export Inspection Council (EIC) acts as a facilitator and intermediary between the importing countries and export units. Therefore, any notification regarding Maximum Residual Limit (MRL), SPS, PPM, and Labelling and Packaging (L&P) received from importing countries is communicated to these units through EIC which applies to all processing-cum-exporting units uniformly. It also advises the Government on measures to be taken to enforce quality control and inspection for exports. The EIA provides pre- shipment inspection and certification services.

3A.2.1 Domestic Standards

The Ministry of Health and Family Welfare (MOHFW) deals with the programme of food quality and safety at the national level under the ambit of the legislation called Prevention of Food Adulteration Act, 1954 [PFA Act has therefore designated a National Codex Contact Point (NCCP-India) for liaison with the CAC. The NCCP has been functioning since 1971. Under the PFA (Prevention of Food Adulteration Act) Act, 1954, MOHFW is authorized to lay down standards for food processed and marketed domestically. It is an act to make provision for prevention of adulteration of the food.

a) This Act extends to the whole of India,[1]
b) It shall come into force on such date as the Central Government may, by notification in the Official Gazette, appoint.

The standards laid down under the above relate to the following areas:

1. Packaging and labelling,
2. Poisonous metals,
3. Flavouring agents and related substances, and
4. Insecticides and pesticides.

a) Instructions by Notification. No. GSR 754 (E) dated 15.5.1976 (w.e.f. 1.7.1976) and Instructions by Notification. No. GSR 790 (E) dated 10.10.1983:

Restriction on the use of insecticides, subject to the Provisions of the Sub rule:

Sub rule:

The amount of insecticides in the food shall not exceed the tolerance limit prescribed:

Table 3A.1 Prescribed Tolerance Limit of Certain Insecticides found in Fish and Fishery Products in India

Name of insecticide	Food	Tolerance limit mg/kg (ppm)
Carbaryl	Fish (Ins. by Noti. No. GSR 174 [E] dated 6.4.1998).	0.2
D.D.T (the limits apply to D.D.T., D.D.D and D.D.E. singly or in any combination)	Fish	7.0
Endosulfan (residues are measured and reported as total of endosulfan A and B and endosulfan-sulphate)	Fish (Amended by Noti GSR 174 [E] dated 6.4.1998)	0.20
Fenitrothion	Fish	0.25
Delta (™) isomer:	Fish	0.25
Hexachlorocyclohexane and its isomers (a) Alfa (⟨) isomer:	Fish	0.25
(b) Beta (β) isomer	Fish (Amended Vide No. GSR 174 [E] dated 6.4.98)	0.25
Quinolphos	Fish	0.01

Source: www.mohfw.nic.in

5. Antibiotics and other Pharmacologically Active Substances:

Ins GSR 771 (E) dated 29.9.2003- and GSR 322 (E) under the Prevention of Food Adulteration (… Amendment) Rules 2003—Residues of antibiotic and other pharmacologically Active Substances.

The amount of antibiotic mentioned in column (1) in the sea foods including shrimps, prawns or any other variety of fish and fishery products, shall not exceed the tolerance limit prescribed in column (2) of the table given below:

Table 3A.2 Prescribed Tolerance Limit of Certain Antibiotics found in Fish and Fishery Products in India

Name of Antibiotics	Tolerance limit mg/kg *(ppm)*
Tetracycline	0.1
Oxytetracycline	0.1
Trimethoprim	0.05
Oxolinic acid	0.3

Source: www.mohfw.nic.in

(i) All Nitrofurans including
    (a) Furaltadone
    (b) Furazolidon
    (c) Furylfuramide
    (d) Nifuratel
    (e) Nifuroxime
    (f) Nifurprazine
    (g) Nitrofurnatoin
    (h) Nitrofurazone
(ii) Chloramphenicol
(iii) Neomycin
(iv) Nalidixic acid
(v) Sulphamethoxazole
(vi) Aristolochia spp and preparations thereof
(vii) Chloroform
(viii) Cholropromaszine
(ix) Colchicine
(x) Dapsone
(xi) Dimetridazole
(xii) Metronidazole
(xiii) Ronidazole
(xiv) Ipronidazole
(xv) Other nitromidazoles
(xvi) Clenbuterol
(xvii) Diethylstibestrol (DES)
(xviii) Sulfanoamide drugs (except approved Sulfadimethoxine, Sulfabromomethazine, and Sulfaethoxypyridazine)
(xix) Fluoroquinolones
(xx) Glycopeptides

3A.3 CONSTRUCTION OF THE INDEX OF NON-TARIFF MEASURES

An index of NTMs for EU, the US, and Japan was constructed with the objective of estimating the impact of existing food safety standards on the volume of exports from West Bengal, India. It is believed that these standards act as an obstacle to free trade of food products, including fish and fishery products and are discriminatory in nature.

3A.4 SOURCES AND VARIABLES

Data on five major variables (GS – General Stringency, MRL – Maximum Residual Limit, L&P – Labelling and Packaging, PPM – Product and Process Method, and SPS – Sanitary and Phyto-Sanitary Measures), constituting NTMs was obtained from TRAINS CD-ROM and various official websites of the three countries.

3A.5 METHODOLOGY

3A.5.1 Ranking of the Variables for Non-tariff Measures

To capture the impact of NTMs on the volume and trend of trade over time, an aggregation of numerous regulations into one index is required. These regulations are in the form of food safety standards imposed by the developed countries on the export of fish and fishery products from the developing countries. As mentioned earlier, these standards are seen as barriers to free trade of food products. In other words, they are considered as non-tariff measures imposed by the developed countries on their imports from other countries, particularly developing countries. And the countries exporting fish and fishery products, as in our case, have to adhere to these standards. The present index of such measures is based on the method of ranking of stringency over time with higher weights for increasing stringency as per the method (Kadekodi and Mishra 2003). It is assumed that the stringency of a particular measure increases equi-proportionately with each successive amendment of that measure.

3A.5.2 Index of Non-tariff Measures

An aggregated index has been constructed for each country using the Factor Analysis Method on the basis of the ranks assigned to each component of the NTMs.

GS: The level of General Stringency is given numerical values ranging from 1 to 7 on the basis of the following regulations:

Table 3A.3: Ranking of Non-tariff Measures Imposed by the European Union on its Fish Imports

Year	GS	MRL	L &P	PPM	SPS
1989	1	1	1	1	1
1990	1	1	1	1	1
1991	2	1	1	1	1
1992	2	1	2	2	2
1993	3	2	2	2	3
1994	3	2	2	2	3
1995	4	3	2	3	4
1996	4	3	2	3	4
1997	5	3	2	4	4
1998	5	3	2	4	4
1999	5	3	2	4	4
2000	5	3	3	4	4
2001	5	4	4	4	4
2002	6	4	4	4	5
2003	7	5	5	5	6

Source: Data taken from the relevant official websites and converted into numbers.

1. Rules and regulations regarding food safety, hygiene, and sanitation existed and the same were applied on fish imports.
2. Need for verifying and certifying compliance of Fishery and Aquaculture Products with the requirements of Directive No 91/493/EEC.
3. EC Directive (93/43) lays down general rules of hygiene for foodstuffs and procedures for verification and compliance with these rules.
4. Quality Control, Inspection, and Monitoring Rules, 1995 on imports of Fresh, Frozen, and Processed Fish and Fishery Products was applied.
5. On coming into force of EC Directive (No.97/876/EC) requiring units to get approval certificate from EIC.
6. On coming into force the regulations (178/2002/EC) laying down general principles and requirements of food law, establishing the European Food Safety Authority.
7. Amendment of decision 95/408/EC of 22 June 1995 (2003/204/EC) on the conditions for drawing up provisional lists of third country

establishments from which Member States are authorized to import fishery products or live bi-valve molluscs.

MRL: Stringency under the Maximum Residue Limits is given numerical values ranging from 1 to 4 on the basis of the following regulations:

1. EC recommended (Doc. VI/1971) levels for residual chlorine levels in the water used for washing fishery products. All units exporting these products to EU have to meet this limit.
2. Decision 93/351 concerning maximum levels of mercury was taken and the same was implemented on fish and fishery products imports.
3. Revision of EC recommendation (Doc. VI/1971) in 1995 for residual chlorine level not above 2 ppm in the water used for washing fishery products. EU adopted acceptable level of chlorophenicol in shrimp (0.3 ppm)
4. In 2001, EU initiated a food safety policy called 'zero tolerance' towards chloramphenicol, nitrofuran, and other antibiotics in shrimp. The residue in shrimp should be between 0.3 ppb to 0.7 ppb.
5. Recommendation 2003/10/EC concerning a coordinated programme for the official control of foodstuffs for 2003. Bacteriological safety and level of histamine for certain fish species was also reported.

L&P: Stringency under the Labelling and Packaging measures is given numerical values ranging from 1 to 5 on the basis of the following regulations:

1. Packaging standards must be in accordance with Directive 79/112/EEC and Council Directive 89/396/EEC of 14 June, 1989.
2. Application of Annex II to Directive 92/48/EEC specifying packaging requirements.
3. Regulation 104/2000 regarding labelling information for the consumers concerning fishery and acquaculture products came into force.
4. Regulation 104/2000 regarding labelling information for consumers of fishery and aquaculture products amended in 2001 implying stringent regulations on imports from other countries.
5. Article 6 of Decision 2003/858/EC concerning packaging of fish products of aquaculture origin to be applied on the imports of such products.

PPM: Stringency under the Product Processing Methods is given numerical values ranging from 1 to 5 on the basis of the following regulations:

1. Processing units are required to water quality test (EEC Directive 80/778/EEC).

70 Biodiversity, Land-use Change, and Human Well-being

2. On coming into effect of Annexure I and II of EC directive 92/48 requiring freezing vessels to meet storage and transport standards.
3. Clause 13 of Notification [S.O.730 (E)] dated August 21, 1995 authorizing EIAs for ensuring compliance by the establishments/units exporting fishery products to the EU Countries.
4. EC Directive (No.97/876/EC) requiring units to get approval certificate from EIC.
5. On application of Article 6 of Decision 2003/858/EC concerning techniques used in processing fish products of aquaculture origin.

SPS: Stringency under the Sanitary and Phyto-sanitary measures is given numerical values ranging from 1 to 6 on the basis of the following regulations:

1. Council Directive (91/493/EEC) lays down the health conditions for the production and the placing on the market of fishery products.
2. Directive 92/48/EEC concerning minimum hygiene rules applicable to fishery products was also introduced.
3. EC Directive (93/43) lays down general rules of hygiene for foodstuffs and procedures for verification and compliance with these rules became mandatory.
4. Directive 95/71/EC amendment of 91/493/EEC directive came into force.
5. Regulations (178/2002/EC) laying down general principles and requirements of food law, establishing the European Food Safety Authority were imposed.
6. Annex V of Decision (2003/858/EC) requiring certificate to be attached with the consignments that are exported to the EU Countries must show that acquaculture products are free of clinical signs of disease. Also the certificate must certify that the fish comes from an area with at least the same animal health status.

Table 3A.4: Ranking of Non-tariff Measures Imposed by Japan on its Fish Imports

Year	GS	MRL	L &P	SPS
1989	1	1	1	1
1990	1	1	1	1
1991	1	1	1	1
1992	1	1	1	1
1993	1	1	1	1

(Contd.)

(Table 3A.4 contd.)

Year	GS	MRL	L &P	SPS
1994	1	1	1	1
1995	1	1	1	1
1996	1	1	1	1
1997	1	1	1	1
1998	1	1	1	1
1999	2	2	1	2
2000	2	3	2	2
2001	3	3	3	3
2002	3	3	3	3
2003	4	4	4	4

Source: Data taken from *www.jetro.go.jp* and *www.mhlw.go.jp* and converted into numbers.

GS: The level of General Stringency is given numerical values ranging from 1 to 4 on the basis of the following regulations:

1. Japan enacted its first law (Food Sanitation Law) regarding human health and safety way back in 1947.
2. Japan amended the Food Sanitation Law in 1999 with implications for fish and fishery products that were imported into Japan.
3. In 2001, Specifications and standards were established as preventive action against food poisoning caused by Vibrio parahaemolyticus in fish and shellfish.
4. Notification No. 301 of the Ministry of Health, Labour, and Welfare (MHLW), 2003, concerning the monitoring and guidance of imported foods, additives, equipment, containers, and packages by the national government.

MRL: Stringency under Maximum Residue Limits is given numerical values ranging from 1 to 4 on the basis of the following regulations:

1. Since 1985, the Ministry of Health, Labour, and Welfare in Japan has been conducting surveys of residues, including pesticides to obtain basic data for the establishment of MRLs.
2. The latest edition of official compilation of food additives was done in order to collect specifications and standards for food additives, 1999.
3. In the year 2000 MRLs were established for 199 pesticides on about 130 commodities including aquaculture products.
4. In October 2003, MHLW announced the introduction of a positive

list system for MRLs in foods. All products sold in Japan, including imports, had to comply with these MRLs.

L&P: Stringency under Labelling and Packaging measures is given numerical values ranging from 1 to 4 on the basis of the following regulations:

1. JAS Law, 1970: Quality Labelling Standard System introduced as amendment to JAS Law 1950.
2. Revision in the JAS Law in the year 2000.
3. The Japanese Agricultural Standards (JAS) requiring a 'Clean Fish' mark on all shipments of seafood to Japan, 2001.
4. New 'Package Recycling Law' came into force that required paper and plastic packaging to be appropriately labelled and recycled. It became fully mandatory on April 1, 2003 for all the importers of food products.

SPS: Stringency under Sanitary and Phyto-sanitary measures is given numerical values ranging from 1 to 4 on the basis of the following regulations:

1. A sanitation (health) certificate including information on species of fish, sea area of fishing, etc., which is issued by an official organization of the exporting country became mandatory (E.V. Notice No. 7 of March 3, 1984) for all the fish exporting countries.
2. Amendment in Food Sanitation Law was made in 1999.
3. Specifications and standards were established as preventive action against food poisoning caused by Vibrio parahaemolyticus.
4. Amendment in Food Safety Basic law (Law No. 48) ensuring food safety must be achieved by taking appropriate measures at each step of the food supply process. Basically looking at the production processes. Extensive checking of the food sanitation conditions of various imported foods was carried out.

Table 3A.5: Ranking of Non-tariff Measures Imposed by the United States on its Fish Imports

Year	GS	MRL	L&P	PPM*	SPS
1989	1	1	1	1	1
1990	1	1	2	1	1
1991	1	1	2	1	1
1992	2	1	2	2	1
1993	2	2	2	2	1
1994	2	2	2	2	2

(Contd.)

(Table 3A.5 contd.)

Year	GS	MRL	L&P	PPM*	SPS
1995	2	2	2	3	2
1996	3	2	2	3	2
1997	4	2	3	4	3
1998	4	2	3	5	3
1999	4	2	3	5	3
2000	5	3	3	5	3
2001	5	4	3	5	3
2002	6	5	4	5	3
2003	7	5	5	5	3

Notes: * Product and Process Method.
Source: Data taken from *www.fda.gov* and converted into numbers.

GS: The level of General Stringency is given numerical values ranging from 1 to 7 on the basis of the following regulations:

1. The Public Health Service Act (PHSA: 42 U.S.C. 262, 294 et seq.) ensuring that seafood shipped or received in interstate commerce is 'safe, wholesome, and not misbranded or deceptively packaged' to control the spread of communicable disease (FDA, 1988d) was enforced.
2. Preventive Control Survey, 1992 to establish whether firms exporting their products to the US were employing adequate preventive controls became mandatory.
3. Food Quality Protection Act, 1996 came into force.
4. Hazard Analysis Critical Control Point (HACCP) regulations (21 CFR 123) 1997 for preventing food safety hazards occurring in the finished products were imposed on all imported products.
5. Action Levels for Poisonous and Deleterious Substances in Human Food and Animal Feed were laid down by FDA 2000.
6. Bioterrorism Act 2002 emphasizing protection of US citizens from dangerous food products came into force.
7. Two new FDA rules were issued by the Department of Health and Human Services (HHS) to enhance the security and safety of food supply.

MRL: Stringency under Maximum Residual Limit is given numerical values ranging from 1 to 5 on the basis of the following regulations:

1. Pesticide Monitoring Improvements Act was implemented 1988.
2. In the year 1993, FDA proposed a 'threshold of regulation' policy regarding trivial amounts of toxins in food.

3. 1 ppm of methyl mercury in crustaceans (Action Levels for Poisonous and Deleterious Substances in Human Food and Animal Feed, FDA 2000) was prescribed as the threshold limit.
4. FDA and EPA safety levels of toxic elements as given in Regulations and Guidance 2001 for crustaceans. All units exporting their products to the US had to adhere to these levels.
5. USFDA announced on 14 June 2002 that it would increase the sampling of imported shrimp for the presence of chloromphenicol antibiotic upto 1 ppb.

L&P: Stringency under Labelling and Packaging measures is given numerical values ranging from 1 to 5 on the basis of the following regulations:

1. In 1966 Fair Packaging and Labelling Act came into force.
2. Nutritional Labelling and Education Act came into force in 1990.
3. 21 Code of Federal Regulations Part 101 were laid down by the USFDA (Food Labelling FDA 1997).
4. In 2002 all imports were required to adhere to the Farm Security and Rural Investment Act of 2002. The act required country of origin labelling for fishery products on all fish exports.
5. Bio-terrorism Preparedness and Response Act came into force in 2002.

PPM: Stringency under Product Processing Methods is given numerical values ranging from 1 to 5 on the basis of the following regulations:

1. Product specifications of Non-standardized product, which do not meet US Standards for Grades for Fishery Products, 1989 was specified and made mandatory.
2. June 1992, The US General Accounting Office recommended revamping the federal food safety system making it even difficult to export to the US.
3. FDA Federal Register Rule of January 28, 1994 (59 FR 4142) establishing requirements for processing and importing seafood in the US was implemented.
4. Hazard Analysis Critical Control Point (HACCP) Regulations (21 CFR 123) 1997 for preventing food safety hazards occurring in the finished products were laid down.
5. A new Act called 'Antimicrobial Regulation Technical Corrections Act' of 1998 (ARTCA) (Public Law 105–324), regarding the use of an acidified solution of sodium chlorite used as an antimicrobial agent in water and ice that are used to rinse, wash, thaw, transport, or store seafood is subject to regulation by FDA as a food additive.

SPS: Stringency under the Sanitary and Phyto-sanitary measures is given numerical values ranging from 1 to 3 on the basis of the following regulations:

1. The Public Health Service Act (PHSA: 42 U.S.C. 262, 294 et seq.) ensuring that seafood shipped or received in interstate commerce is 'safe, wholesome, and not misbranded or deceptively packaged' to control the spread of communicable disease (FDA, 1988d) came into force.
2. FDA Federal Register 60 FR 65095 December 18, 1995 defined the procedures for the Safe and Sanitary Processing and Importing of Fish and Fishery Products; Final Rule 21 CFR 123 and 1240.
3. Hazard Analysis Critical Control Point (HACCP) regulations were laid down in 1997.

Table 3A.6: Country Specific Aggregate Index of Non-tariff Measures

Year	EU	US	JPN
1989	1.02	1.06	1.02
1990	1.02	1.27	1.02
1991	1.23	1.27	1.02
1992	1.84	1.70	1.02
1993	2.46	1.90	1.02
1994	2.66	2.54	1.53
1995	3.48	2.96	1.53
1996	3.48	3.18	1.53
1997	3.89	4.04	1.53
1998	3.89	4.25	1.53
1999	3.89	4.25	2.29
2000	4.10	4.67	2.81
2001	4.51	4.86	3.57
2002	4.92	5.50	3.83
2003	5.94	5.93	4.85

Source: Authors' calculations.

APPENDIX 3B

DATA TABLES

Table 3B.1: Major Country-wise Average Unit Value Realization of Frozen Shrimp Exported from India (1997–8 to 2001–2)

(in Rs/Kg)

Countries	Year				
	1997–8	1998–9	1999–2000	2000–1	2001–2
Japan	356.73	416.65	390.87	470.49	389.38
USA	232.17	238.78	283.82	378.79	360.15
UK	205.74	231.06	250.71	290.90	296.59
China	213.39	216.80	275.85	251.86	221.62
France	155.86	196.97	223.10	257.23	287.33
Germany	221.64	231.95	275.29	286.46	309.88
Kuwait	242.94	–	–	644.06	–
Saudi Arabia	150.82	–	–	169.30	171.66

Source: www.indiastat.com

Table 3B.2a: Item-wise Exports of Marine Products from India

Items		1991–2	1992–3	1993–4	1994–5	1995–6
Frozen Shrimp	Q	76,151.00	74,051.00	86,541.00	101,751.00	95,724.00
	V	979.10	1,176.83	1,770.73	2,510.27	2,356.81
Frozen Fish	Q	49,333.00	75,794.00	94,022.00	1,22529.00	100,093.00
	V	143.20	233.53	296.00	446.57	372.26
Fr. Squid	Q	25,529.00	30,364.00	34,741.00	37,197.00	45,025.00
	V	109.40	151.90	192.47	245.10	319.58
Frozen Cuttlefish/Fillet	Q	12,437.00	18,981.00	18,998.00	28,145.00	33,845.00
	V	60.90	118.88	138.18	224.01	260.86
Others	Q	8,370.00	9,835.00	9,658.00	1,7715.00	21,590.00
	V	83.30	87.42	106.24	149.32	191.06
Total	Q	171,820.00	2,09025.00	243,960.00	307,337.00	296,277.00
	V	1,375.89	1,768.56	503.62	3,575.27	3,501.11

Source: Marine Products Export Review, MPEDA (various issues).
Notes: Q: Quantity in M. Tons
V: Value in Rs Crore

Table 3B.2b: Item-wise Exports of Marine Products from India

Items		1996–7	1997–8	1998–9	1999–2000	2000–1	2001–2
Frozen Shrimp	Q	105,426.00	101,318.00	102,484.00	110,275.00	111,874.00	127,709.00
	V	2,701.78	3,140.56	3,344.91	3,645.22	4,481.51	4,139.92
Frozen Fish	Q	173,005.00	188.29	108,556.00	131,304.00	212,903.00	174,976.00
	V	636.92	726.73	495.03	537.34	874.68	713.11
Fr. Squid	Q	40,924.00	35,095.00	32,254.00	34,918.00	37,628.00	39,790.00
	V	290.45	270.89	268.93	296.80	324.43	329.67
Frozen Cuttlefish	Q	31,178.00	3,7258.00	34,589.00	32,799.00	33,677.00	30,568.00
	V	272.37	323.41	273.31	286.22	288.99	280.07
Others	Q	27,666.00	2,4118.00	25,051.00	33,735.00	44,391.00	51,427.00
	V	219.84	235.89	244.69	351.09	474.28	494.28
Total	Q	378,199.00	385,818.00	302,934.00	343,031.00	440,473.00	424,470.00
	V	4,121.36	4,697.48	4,626.87	5,116.67	6,443.89	5,957.05

Source: *Marine Products Export Review*, MPEDA.
Notes: Q: Quantity in M. Tons
V: Value in Rs Crore

Table: 3B.2c: Item-wise Exports of Marine Products from India

Items		2002–3	2003–4
Frozen Shrimp	Q	134,815.00	129,768.00
	V	4,608.31	4,013.07
Frozen Fish	Q	196,322.00	1,38023.00
	V	841.65	620.73
Fr. Squid	Q	37,838.00	37,832.00
	V	384.37	372.92
Frozen Cuttlefish	Q	41,381.00	39,610.00
	V	417.09	435.18
Dried Items	Q	8,178.00	12,574.00
	V	84.23	145.68
Live Items	Q	2,115.00	2,341.00
	V	53.66	51.10
Chilled Items	Q	3,350.00	3,779.00
	V	59.14	64.04
Others	Q	43,298.00	48,090.00
	V	432.86	389.23
Total	Q	467,297.00	412,017.00
	V	6,881.31	6,091.95

Source: *Marine Products Export Review*, MPEDA.
Notes: Q: Quantity in M. Tons
V: Value in Rs Crore

Appendix 3C

COUNTRY-WISE TRENDS IN THE INDEX OF NON-TARIFF MEASURES AND REAL VALUE OF EXPORTS

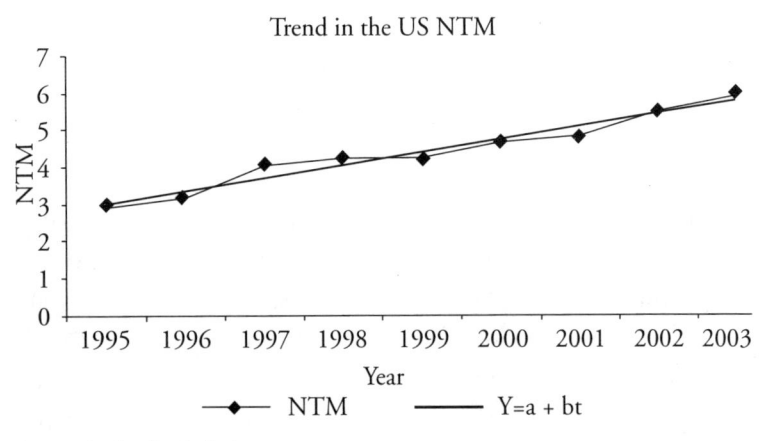

Source: Authors' calculations.

Figure 3C.1: Trend in the Index of Non-tariff Measures specific to the US

Source: Authors' calculations.

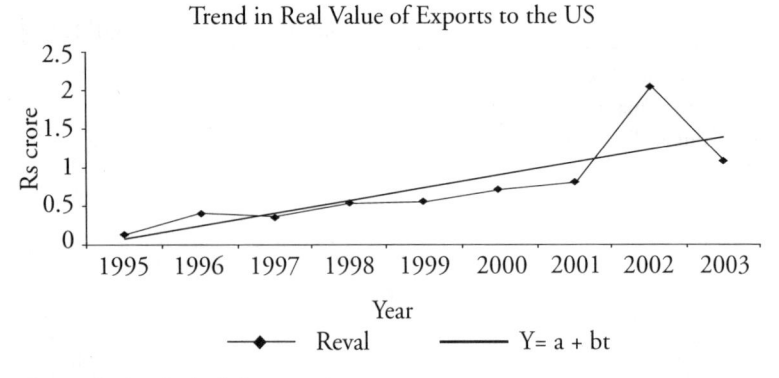

Figure 3C.2: Trend in the Real Value of Frozen Shrimp Exports to the US

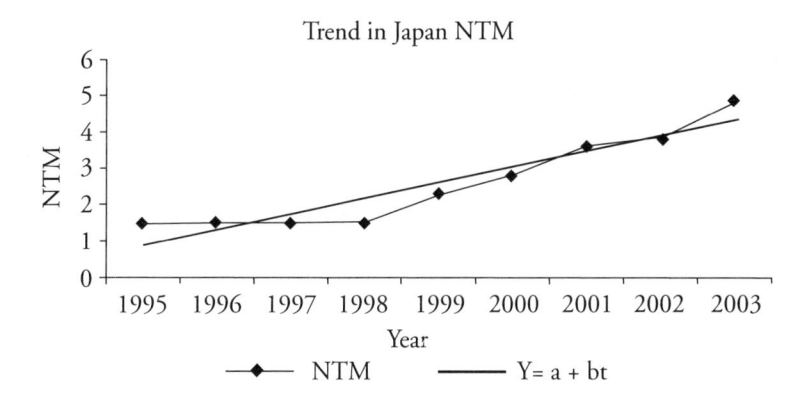

Trend in Japan NTM

Source: Authors' calculations.

Figure 3C.3: Trend in the Index of Non-tariff Measures Specific to Japan

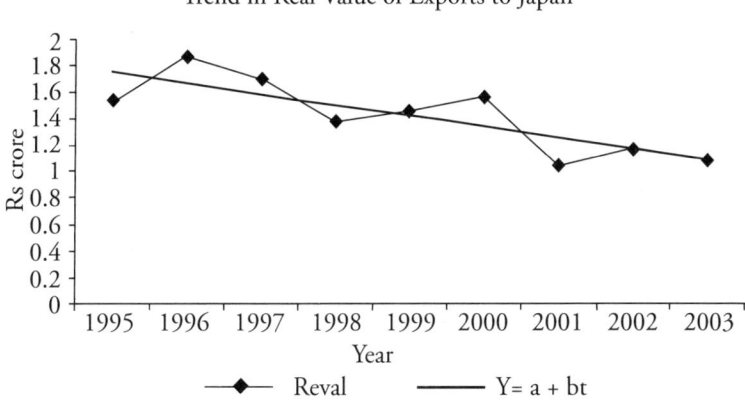

Trend in Real Value of Exports to Japan

Source: Authors' calculations.

Figure 3C.4: Trend in the Real Value of Frozen Shrimp Exports to Japan

Trend in UK NTM

Source: Authors' calculations.

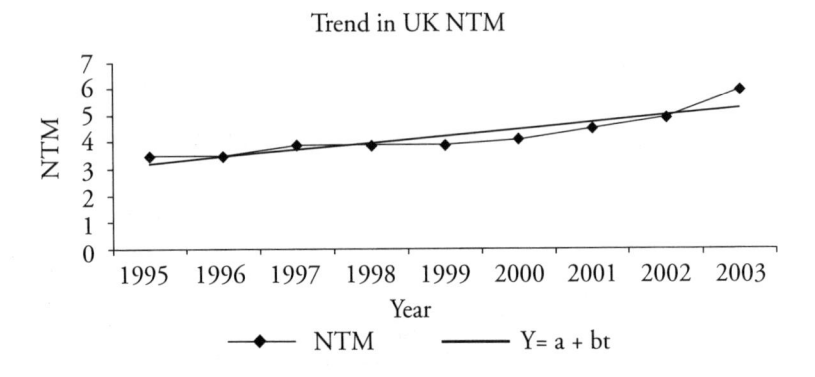

Figure 3C.5: Trend in the Index of Non-tariff Measures Specific to UK

Trend in Real Value of Exports to UK

Source: Authors' calculations.

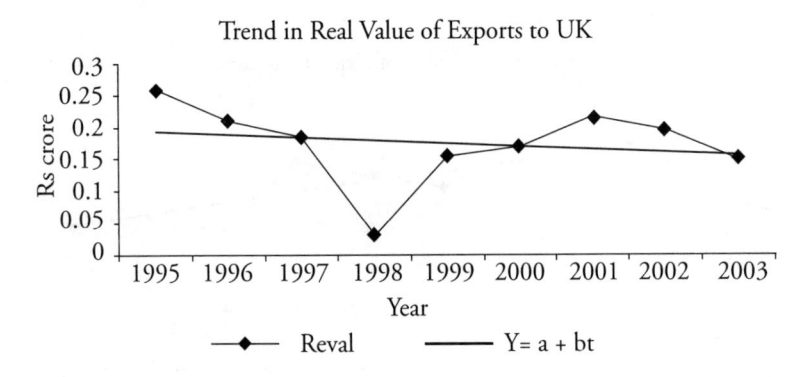

Figure 3C.6: Trend in the Real Value of Frozen Shrimp Exports to UK

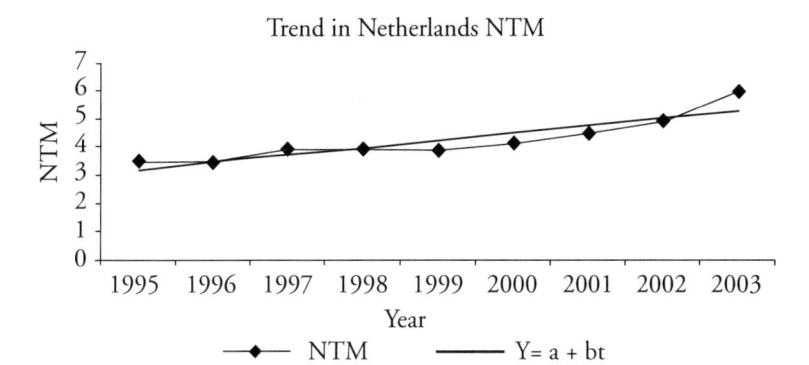

Source: Authors' calculations.

Figure 3C.7: Trend in the Index of Non-tariff Measures Specific to Netherlands

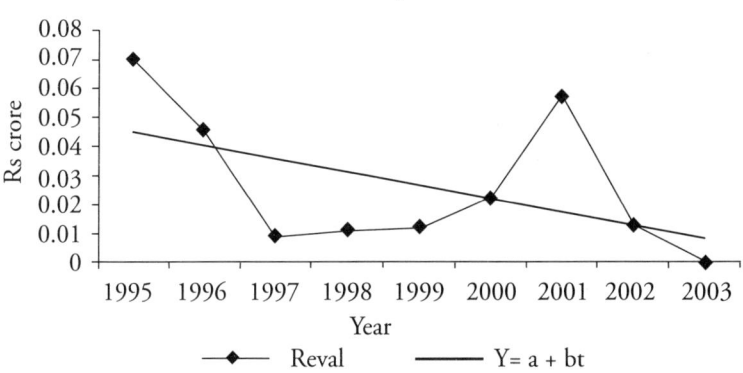

Source: Authors' calculations.

Figure 3C.8: Trend in the Real Value of Frozen Shrimp Exports to Netherlands

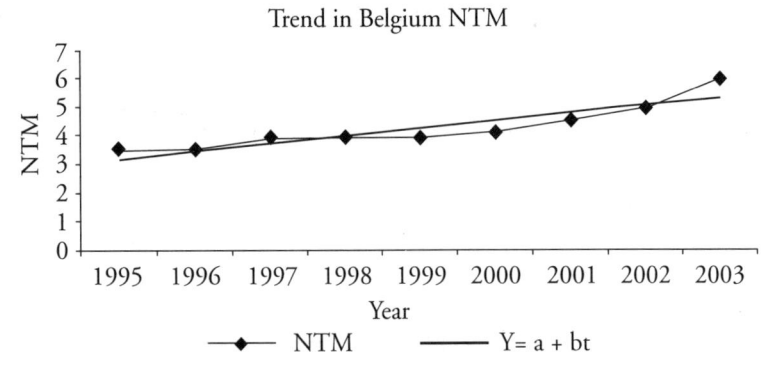

84 Biodiversity, Land-use Change, and Human Well-being

Trend in Belgium NTM

Source: Authors' calculations.

Figure 3C.9: Trend in the Index of Non-tariff Measures Specific to Belgium

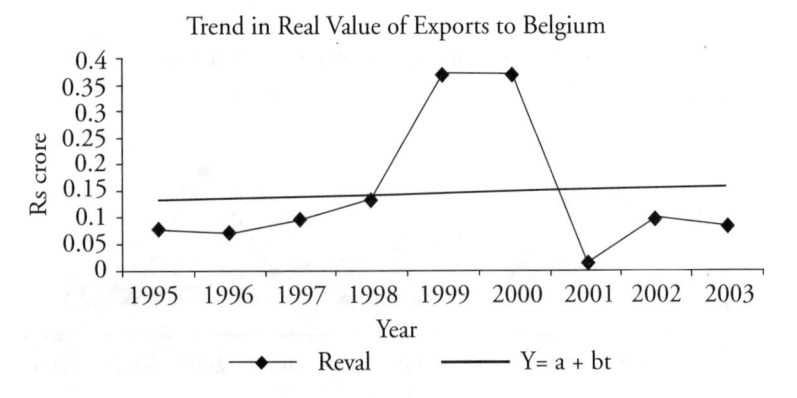

Trend in Real Value of Exports to Belgium

Source: Authors' calculations.

Figure 3C.10: Trend in the Real Value of Frozen Shrimp Exports to Belgium

NOTES

1. The word 'except the state of Jammu and Kashmir' omitted by Act 41 of 1971, s. 2 (w.e.f. 26.1.1972).

4

Shrimp Exports and Aquaculture
The Region and the Stakeholders

INTRODUCTION

Under the reforms package, tariffs and quantitative restrictions were reduced and private initiatives in shrimp culture, shrimp processing, and its exports were encouraged. Due to the increased market linkages, in West Bengal, as elsewhere, shrimp culture activity was transformed from a subsistence-oriented to a market-driven economic activity. Commercial shrimp farming generated employment opportunities for people living in the coastal regions. Also, it has been a major source of foreign exchange earnings besides being a source of earning for people who are involved in it directly as well indirectly. From the *meen* (or the post-larvae (PL)) collector to the final processing unit, there is an entire chain of players who have significant roles to play in the production and export of shrimp. All of them constitute the set of stakeholders in it.

This increased processing and export of shrimp from West Bengal initiated a number of significant changes in the economy of the region. The activity also resulted in the emergence of a large number of stakeholders, spread in the region. All of them were directly or indirectly involved in the production, transport, processing, and export of shrimp. These changes in the region had different dimensions over space and time:

1. Its impact on incomes, livelihoods and well-being of a range of stakeholders,

2. Its impact on the use of natural resources such as land and water and larger consequences for biodiversity in the region, and
3. Finally, in the medium and long run, there could be a feedback effect of the changing resource base on livelihood and well-being of people in the region.

These effects acquire significant dimensions in view of the ecological significance of the region and the differential socio-economic characteristics of the stakeholders. This chapter describes these important aspects of the region and the stakeholders.

It is important to point out here that spatial units carry different connotations from the economic, social, and ecological perspectives. Most market-driven economic activities are viewed as taking place within or across administrative regions. Bio-physical characteristics and socio-cultural aspects of spatial entities are secondary and seen as significant only when they either promote or hinder economic activity. Most of the time distance is seen as an obstacle to be overcome by better market integration and improvements in physical and social infrastructure. In studying the ecological impacts of economic activity, the bio-physical characteristics of spatial entities are central. This study examines the inter-relation between increased aquaculture and regional impacts on human well-being, land-use, and off-shore biodiversity. Our approach to space is therefore a pragmatically driven one. It is the means as well as the vehicle through which economic activity leads to human well-being. We shall not only focus on the Sundarbans as a region but also switch back and forth to administrative units and areas through which the well-integrated market chains of stakeholders operate in reaching out to the processing units who are the exporters.

THE SUNDARBANS REGION—BACKGROUND AND OVERVIEW

The Sundarbans delta spanning 355 sq km in width is the largest mangrove forest in the world at the mouth of the Ganges. Up to the year 1770, the total area of Sundarbans of India and Bangladesh was estimated to be around 36,000 sq km, which at present, stands at 25,000 sq km. The Indian part consists of 9630 sq km. The rest lies within Bangladesh. Out of the 9630 sq km, 4264 sq km of

Shrimp Exports and Aquaculture 87

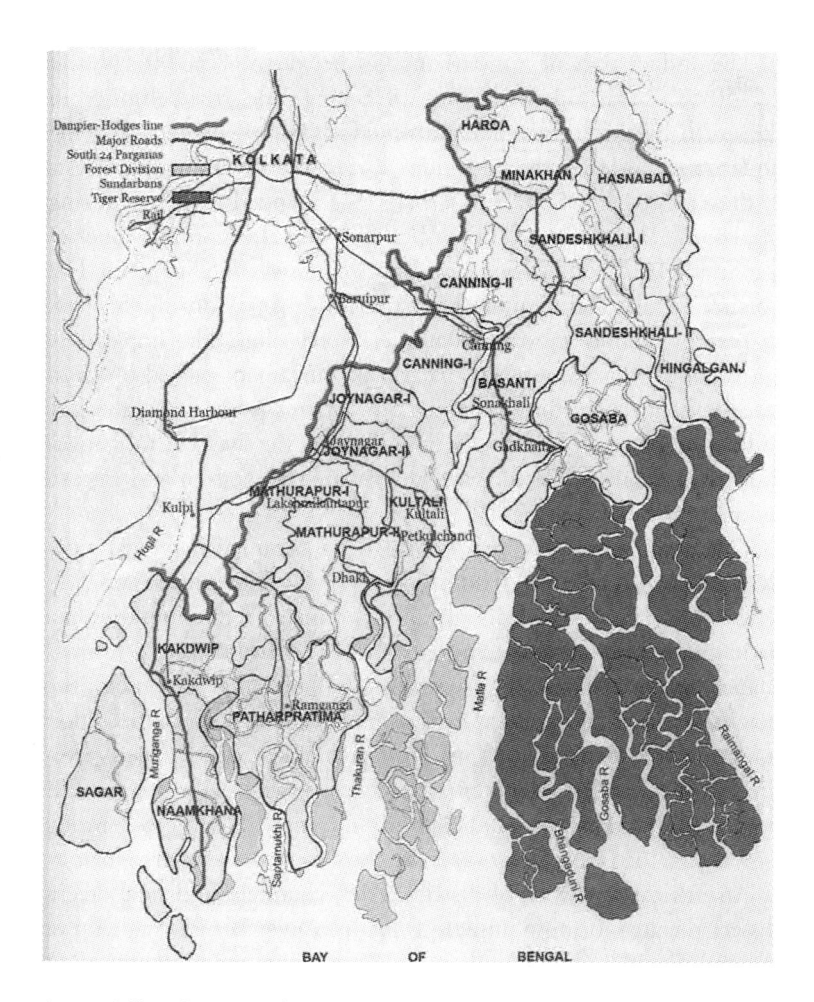

Source: Office of WWF, India

Map 4.1:   Sundarbans in West Bengal

wetland/mangroves constitutes reserve forests, which in turn comprises of 2195 sq km of wetland-mangroves and 2069 sq km of tidal river. This means that the reclaimed area of around 5,366 sq km forms human settlements in 19 blocks (13 in district of South 24 Paraganas and 6 in the district of North 24 Paraganas).

The Indian part of Sundarbans lies between 21°30'00' 'N and 22° 40'48' 'N latitude and 88° 1'48"E longitude. It is delimited in the north by the so-called 'Dampier-Hodges line' demarcating the northern extension of the inter-tidal zone marked by mangrove forests as they existed in 1830. Sundarbans was mapped in detail during the 1830s. At that time William Dampier was the Commissioner of the Sundarban Commission and Lt. Hodges was the Surveyor. The northern limit of the Sundarbans mangrove forest during 1830 as marked by them is known as Dampier-Hodges line. This line closely corresponds to Kakdwip-Basirhat-Dhaka lineament picked up from satellite imagery. In the south, the Sundarbans is bound by the Bay of Bengal. The river Hoogly (in the west) and the river Harinbhanga-Taimanga-Ichamati (in the east) demarcate the western and eastern boundaries respectively.

During recent times, the Bengal delta acquired a typical tide-dominated nature with a tidal range varying between 3.7 to 5m. The estuarine mouth of Hoogly and its numerous distributaries like Saptamukhi, Jamira etc. acquired a typical seaward flaring funnel-shaped pattern. Due to progressive shallowing of channels, the height of the tidal bore acquired a maximum of 6.4m and further inland this increased to 7.17 m. The flood and ebb tides have a semi-diurnal nature (12.5 hrs interval) and occur twice daily. Within this cycle floodwater flows for 2–3 hours duration. For the remaining 8–9 hours, the estuary is covered by ebb tide flow of lesser velocity.

An intricate network of distributaries, channels, and tidal creeks dissect the area forming numerous plano-convex islands made up of silt and silty clay. The islands of 3 to 8m height are partially/ often completely inundated by water during high tide. The subsidence in these areas is comparatively more severe than in the open coastal parts of Digha and these are often manifested by mild earth tremors occasionally. Somewhat more rarely there are instances of sudden subsidence like that during 1897, which created a series of '*Bils*' instantaneously at Rangpur and Mymansingh now in Bangladesh.[1]

The Bengal Basin, and this part of the deltaic plain is gradually tilting towards the east. This has probably caused the main fresh water discharge to shift gradually eastward (through Bangladesh) imposing severe stress on freshwater budget for Hoogly–Matla estuary. The eastern part of the Sundarbans delta is experiencing a higher rate of subsidence due to sediment loading. This subsidence rate is often close to 6mm/ year as measured at some places of Bangladesh.

Sundarbans, the only mangrove tiger-land of the globe is presently under threat of severe coastal erosion due to sea level rise. It appears that the once largest prograding delta, which registers the highest species diversity in terms of mangrove and mangrove-associate flora and fauna, is showing evidences that suggest the rich biodiversity is under threat. Commentators believe that this deltaic island system is facing degradation due to natural and anthropogenic changes. Frequent embankment failures, submergence and flooding, beach erosion and siltation at jetties and navigational channels, cyclone and storm surges are all making this area increasingly vulnerable.

In addition, alarming growth of population in this ecologically sensitive and fragile niche has posed a major threat for its very existence. Wide scale reclamation, deforestation, and unsustainable resource exploitation practices have together produced changes in the physical and biological dynamics of the coastal system. Several changing parameters of the ecosystem listed below which make the Sundarban extremely vulnerable from the perspective of adaptation to climate related changes:

- Rise in temperature
- Change in rainfall pattern
- Increased intensity of cyclones and storm surges leading to flooding/ inundation
- Rise in relative sea level
- Erosion/ submergence of islands
- Increasing breaches and vulnerability of protective embankments
- Increasing salinization of fresh water rivers, soil
- Loss of bio-diversity, species migration
- Deforestation, reclamation, and land-use change
- Increasing population pressure
- Increasing dependence on rain-fed mono-crop agriculture
- Nitrogen and phosphorous loading and pollution in the coastal water
- Decline in fish catch/ unit effort

- Species destruction during prawn seed collection
- Low level of development and lack of infrastructure
- Prevalence of extreme poverty
- Lack of health facilities and high incidence of vector/ water borne diseases
- Lack of administrative coordination and Integrated Management and Development plans.

SOCIO-ECONOMIC PROFILE OF THE SUNDARBANS

Sundarbans, though an economically backward area, has attracted people from other places due to its rich natural resources, offering a very productive environment by way of sustaining a variety of economic activities, based on agriculture, fishery, and forestry. As per the 2001 Census, the total population of the Sundarbans region was about 3.7 million while as per the (1991a) census, the total population of the region was about 3.2 million, and 2.4 million as per 1981 Census, thus containing an additional population of about 0.8 million and 0.5 million respectively within a decade from 1981–91 and 1991–2001 respectively. The decennial growth rates registered during 1971–81, 1981–91 and 1991–2001 were 21.47 per cent, 29.55 per cent, and 17.4 per cent respectively. Recently the population has touched the figure of 4.2 million, which is predominated by farming and fishing community. The density of population is 845.10 persons sq km with a high concentration of backward classes in the region. Scheduled castes comprise 39.04 per cent while scheduled tribes represent 5.06 per cent of the total population in the region.

Agriculture plays an important role in the local economy by supporting about 89 per cent of the total population of this region as against about 57 per cent of the State. Rice is the dominant food crop of the region and is grown predominantly with potato. The greater part of the region is mono-cropped with only a negligible portion under the second crop. 13 per cent of the total population of the region works as agricultural workers as against 9 per cent in the State. More than 90 per cent of the farming community is either small or marginal farmers and about 50 per cent survives below the poverty level. These poor people work on daily wages in the fields during the monsoon season and then get involved in the harvesting of paddy.[2] For the rest of the year, these labourers remain jobless with a small percentage of them seeking jobs in the fish/shrimp cultivation or fish markets. The poorest of the lot are forced to feed themselves through collection of shrimp seed from the wild. Sundarbans exhibit paddy-cum-fishery system due to the occasional inundation of paddy fields by brackish water. This system has developed into large-scale commercial fisheries by controlled inundation of large tracts of paddy land enclosed by ring bunds.

Next to agriculture, fisheries provide a distinct source of employment and income for the people particularly the small and marginal farmers. Sundarbans has around 4.78 lakh persons engaged in the fisheries activities. Besides, various forms of village arts and crafts are also practiced in the region. Apart from brick making, oil and rice milling is common in most villages. Bamboo net weaving, fishing net weaving, and palm leaf handicrafts such as hats, bags, and other luxury items are also common. Local females in the villages are involved in small tailoring units and handicrafts making, bee keeping and processing of honey etc. is gradually increasing in the region.

Literacy levels and civic amenities available to the population of the Sundarbans present a mixed picture. According to the 2001 Census, the literacy rate in the Sundarbans (54 per cent) was higher than the States literacy rate of 47.2 per cent. However, figures in terms of electrification of households in the Sundarbans and at the State level show that a very small number of households in the Sundarbans—only around 0.042 million (6 per cent) as compared to 5.88 million (37 per cent) in the State are electrified. The situation is not all that grim as far as percentage of households having toilet facilities are concerned. Around 30 per cent of the population in the Sundarbans has access to toilet facilities whereas at the State level the figure is around 44 per cent. Further, the people in the Sundarbans consider handpumps as the most important source of drinking water. Around 0.5 million or 80 per cent of the households get their water from this source. The other sources that are also used for collecting drinking water are tubewells and taps (19 per cent). Finally, the fuel for cooking in the Sundarbans mainly comes from crop residue followed by cowdung and firewood. The percentage of households dependent on these types of fuel is around 90 per cent with a very small percentage (around 1.5 per cent) dependent on coal, lignite, and charcoal. As for the entire state, firewood happens to be the most important type of fuel used for cooking by the households (30 per cent) followed by crop residue (22.31 per cent), cowdung (12.08 per cent) and coal, lignite, and charcoal. Besides these around 13 per cent of the households also use LPG.

The monthly per capita consumption expenditure (based on 30-day recall period) of the households living in North 24 Parganas was around Rs 670 whereas it was Rs 504 for those living in the South 24 Parganas. The corresponding overall figure for the State was close to Rs 579 as per the 55th Round of the National Sample Survey Organisation (NSSO). The figure of Rs 504 is a closer approximation to the actual per capita consumption expenditure in the Sundarbans.

Further, the districts of both North and South 24 Parganas have around 16 per cent of the total hospitals in West Bengal. In absolute terms, this means that 69 hospitals of the total 429 hospitals existed in the State as on 31st March 2002. These hospitals are run by State and Central government, local board, and voluntary organizations. If we look at these two districts separately, we find that South 24 Parganas has just about 5 per cent of the total hospitals in the State and the remaining 11 per cent are in the district of North 24 Parganas. On the other hand, health facilities in terms of availability of Sub-Centres, Primary, and Community health centres are more in South 24 Parganas than in North 24 Parganas. The availability of these centres (as on September 2005) in these two districts was around 10 per cent and 7 per cent respectively. That is, of the total 11,624 health centres in the State as on 30th September 2005, the district of South 24 Parganas had 1,161 centres and the North 24 Parganas even lesser around 817 in all.

It appears that a large part of the socio-economic backwardness of the Sundarbans is related to its physical segregation from the mainland and its specially fragile ecosystem. Lack of adequate and frequent transport facilities to the mainland and the absence of grid-based electric power is visible everywhere. This results in dependence on biomass for fuel and inadequate medical facilities. At the same time, this offers an opportunity to examine the viability of alternative

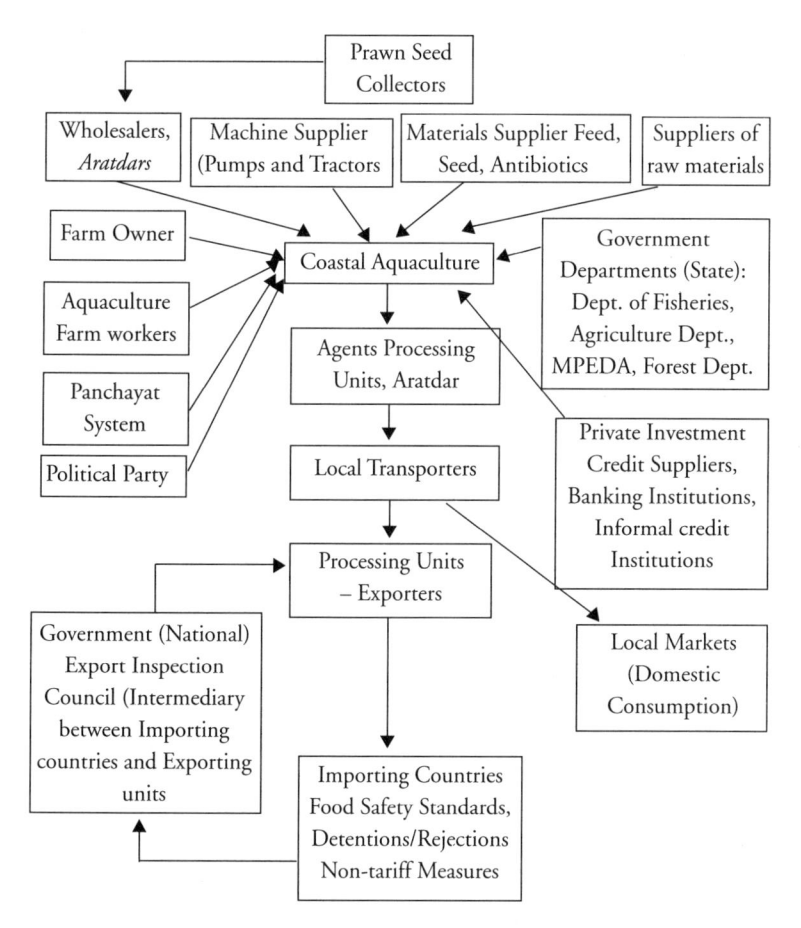

Source: Authors.

Figure 4.1: Different Stakeholders in Aquaculture

sources of electrification, which are being tried on a limited scale. The major strength of the region is the high literacy rates and this needs to be tapped into to design an alternate model of development suitable for the region with its unique ecosystem.

STAKEHOLDERS IN SHRIMP PRODUCTION

The major portion of the coastal zone of Bay of Bengal in the state of West Bengal is covered with rich mangrove vegetation and the brackish-cum-saline aquatic phase of this environment nourishes the world's most famous mangrove forest—the Sundarbans. Surface water is generally saline, giving the Sundarbans a high comparative advantage for various types of brackish water fish production systems including shrimp farming. This activity received a major impetus due to increased exports as documented in earlier chapters. Figure 4.1 shows the stakeholders in this export-driven chain of activities and their location. In the following sections, this chapter attempts an overview of the nature and magnitude of this effect in the region as it impacted different stakeholders.

While the prawn seed collectors concentrated in the coastal region, the network of wholesale traders is spread over the blocks of the Sundarbans, most of the aquaculture is concentrated in Canning, Gosaba, and Minakhan blocks of the two districts and output is transported by a network of transporters to the processing and exporting units located in the vicinity of Kolkata. The following sections constitute in-depth description of the network of stakeholders.

Processing and Export Units

Between 1995–6 and 2004–5, export of frozen shrimp from Kolkata port increased from Rs 354.6 crore to Rs 562 crore.[3] Almost all this was being processed in units in and around Kolkata. While some of these processing units' activities processed other marine products as well, a large number concentrated on shrimp processing and export. Those units which process the entire range of marine products are operational for 10–12 months in the year while those that concentrate on shrimp operate for approximately eight months.

A study of the cost structure of the units reveals that 90–92 per cent of input costs are attributable to raw material cost, that is, purchase of shrimp. Both cultured and captured shrimp are purchased, mainly from agents. Between 80–100 per cent of raw material is purchased through agents. This implies that about Rs 450 to 500[4] crore of shrimp was collected from the hinterland and transported to the units Ca. 2004. This generates incomes for producers, transporters, and others.

These processing units have succeeded in maintaining their competitiveness in foreign markets inspite of increasing stringency in food safety and other standards in all markets, in particular EU markets. This has been possible due to increasing investment in maintenance of standards such as, for example, effluent and water treatment plants. It was observed that there existed an inverse relationship between the scale of operation and cost of compliance with cost varying from 1 to 5 per cent of their total cost of production.

Processing plants fear that the increased cost of compliance with these standards will make them uncompetitive in the export market and affect their profitability of operations, thus driving them out of the competition. With each processing plant offering employment to a number of people, predominantly women, a sizeable number of livelihoods will be lost with their closure. After the EU ban in 1997, the Indian government specified various process standards in order to meet the high quality standards laid down by the EU. As a result, the four-month ban was lifted partially.

The government has been extending considerable support to the processing industry to upgrade the existing systems to meet the international standards. The Marine Products Export Development Authority (MPEDA) is responsible for the promotion and regulation of exports of fish and fish products. It provides financial assistance for the following: acquisition of fish processing machinery, installation of generating sets, establishment of chill room facility, installation of water purification system, and setting up of water effluent treatment plants. Besides, insulated iceboxes are provided at subsidized cost for use in capture and culture operations and interest subsidy is provided for seafood units to facilitate upgradation. Our survey units in West Bengal reported to have obtained financial assistance from MPEDA in setting up of either Effluent Treatment Plants (ETP) or Water Treatment Plants. The subsidy is provided at 25 per cent of ETP cost or Rs 1 lakh, whichever is lower.

Shrimp Farmers

Aquaculture expanded at a rapid rate in West Bengal, particularly in the Sundarbans. All eight blocks in the region registered increases in land under this category, in particular Sandeshkhali, Minakhan, and Namkhana. Minakhan registered the largest conversion of land to aquaculture from 1986 onwards up to 2004.[5]

Table 4.1: Land Converted to Shrimp Culture in the Sundarbans—1986 to 2004

Time Period	Land Converted in sq km
1986–9	107.355
1989–96	143.342
1996–2001	114.891
2001–4	97.646

Source: NRSA satellite data (See Chapter 5)

The categories of stakeholders significant in this context and the issues of significance to them are discussed below.

FARM OWNER-CUM-WORKER

These are aquaculturists who own the farm or have taken it on lease for a short period and work on it simultaneously. About 90 per cent of the farmers surveyed have leased the land and the rest own the land on which they farm. Further, of those who hold a lease on the land, four-fifths have a three-year lease. Only a small minority of 20 per cent has a long-term 20-year lease on the land. All these are located in Canning.

However, a large number of the lease-holders have been farming the same land for more than 8–10 years, some even upto 25 years. It is obvious that the lease is renewed every three years with the option of increasing the area under aquaculture being with the farmers. Aquaculture for these farmers is quite a rewarding and profitable business opportunity. Like other forms of farming, shrimp production involves substantial capital investment and many risks. The owner bears the cost of production and the risk associated with it.

The single most important issue for this set of stakeholders, which determines their profits and their sustainability, is the technology of shrimp production. The prevalent types of culture systems/technologies for shrimp culture in the Sundarbans are as follows:

Traditional System—In this type of system, large areas of water bodies are naturally inundated with water during high tide and require very little management. Water is exchanged infrequently or when required. This system does not require supplementary feed and the production is quite low.

Extensive or Improved Traditional System—This is the most prevalent shrimp culture system in the Indian Sundarbans. Under this system, area of the pond varies between 1–5 hectares with low stocking density (approximately 50,000 seeds/ha). The system requires moderate management while the water is exchanged periodically twice a month during the high tide. Occasionally, feed and fertilizers are used for shrimp growth. This culture system produces reasonable shrimp quantity.

Modified Extensive System—This system employs ponds of 0.5–3 hectares in size with quite high stocking density of around 1 lakh seeds/ha. The system needs proper management of water quality, feed, and good soil characteristics. Water is exchanged quite frequently (6–8 times a month). The pond itself needs suitable preparation before stocking the seeds. Supplementary feed like oil cake etc. are applied whenever required. This system is quite productive and yields about 1.5 tons/ha.

Semi-Intensive System—Under this type of culture system, very small size ponds are used with high quality of management and care. Ponds are prepared by using Mohua oil cakes, medicinal lime etc. Water is exchanged at least thrice a week. The stocking density is very high with an average of properly acclimatized 2 lakh seeds/hectare. The system gives very rich crops with an average production of 4 tons/hectare. But, after the Supreme Court order of 1997, this system has been banned from practice due to its harmful effects on the forest and fish biodiversity in particular; agriculture, and environment in general.

At times a disease outbreak in the culture farm destroys the entire harvest inspite of large amounts spent on medicines and antibiotics to counter the spread of virus. Lack of scientific knowledge of shrimp culture, blindly following what others are doing (using the same medicines and techniques) and no technical assistance by the government are some of the handicaps this activity is facing. Nonetheless, it is a profitable business for a culture farmer.

With expansion in the basket of marine products exports; the demand for other varieties of fish has also increased over the years. To reap the benefits of such an expansion, and also to have an assured income, these people have moved from monoculture to polyculture. The idea being that even if shrimps are attacked by a virus , other species will still survive and fetch them a good price. Though the profit margin in polyculture is less than what it is in monoculture, the farmers feel and in fact have experienced that it is no longer safe to go on with monoculture alone given the vulnerability of shrimps to diseases in the prevailing conditions. There seems to exist a trade-off between stable and high-income accrual.

Farm Workers

The farm-level employment includes two types of workers—temporary low-paid construction workers (pond construction workers) and permanent maintenance labourers (handling, pumping, feeding, pond water treatment, and harvesting), supervisors, and guards to prevent the theft of shrimps from the grow-out ponds. A permanent farm labourer is paid close to Rs 1,300/month with free food and lodging. In some cases, family members of the owner are also involved in aquaculture. Though not paid in cash, they do constitute an important part of the work-force engaged on a farm. The temporary labour is required only for a specific period (few days or may be a month) in a year, mainly for desiltation and construction of the pond, and in case the embankment erodes due to flooding of the pond in the rainy season. Depending on the nature of the work, a temporary labour is paid from about Rs 50/ to Rs 60/day. The aquaculture activity indeed has provided employment and assured income to many poor people in the region.

Prawn Seed Collectors

The supply of hatchery-produced tiger prawn seed *(Penaeus monodon)* in West Bengal is highly inadequate[6] and as a result, the aquaculture farms of this region largely depend on the supply from the wild. Being motivated by a regular cash income, the majority of poor people in the Sundarbans have adopted prawn seed collection as their profession almost throughout the year. The people involved in this activity are from the lowest strata of the society who do not have alternative source of livelihood.[7] At times the entire family is involved in prawn seed collection. Besides prawn seeds, these people also collect crabs, ornamental fish, colour fish, and other varieties of fish from the river or estuary for sale in the local market.

The demand for wild prawn seeds has created considerable employment mostly for women and children in the region. It is estimated that in the Sundarbans about 4 lakh persons including 60,000 children are getting employment and assured income from this activity. They are required to sell the shrimp fries collected during the day to the person (*aratdar*) from whom they have taken (*dadun*) money in advance for meeting their daily needs. The prawn seed collectors on their own cannot sell these seeds in the market at the prevailing prices. Their daily income from the catch varies depending on the number of prawn seeds collected in a day.

Aratdars

The aratdars act as intermediaries between prawn seed collectors and shrimp farm producers/owners. The aratdar is the buyer and seller of the commodity (post larvae in this case). He buys prawn seeds from the PL collectors and sells them to the aquaculture farmer. As a financer of the last resort, the aratdar provides boats and nets to prawn seed collectors on the condition that they will sell their catch to him and also at the same time lends money to an aquaculture farm owner, for meeting his production expenditure. Under this arrangement, the owner agrees to sell his produce to the aratdar.

Input Suppliers (feed, seed, antibiotics, tractors, pumps, aerators, and generators)

Shrimp production involves use of numerous inputs such as shrimp feed, organic and inorganic manures, lime, pesticides and veterinary drugs, as well as technical devices for water treatment and pond preparation. With an increase in the culture activity, the demand for inputs has grown over the years. As a result, supplying of inputs to the owners/producers has become a profitable business for the people living in the coastal areas. Technical devices like, pumps, tractors, aerators, and generators can be hired by paying nominal hourly rental charges. Given the easy accessibility of input suppliers, the owners no more have to invest in these devices. Similarly, with the easy availability of prepared feed from the local market, culture farmers are saved from the difficult task of preparing the feed on their own. All this has led to the development of organized markets in aquaculture farm inputs.

The Political System

Panchayats have a strong presence in the villages where aquaculture activity is carried out and have a role to play. This role can be felt right from the allocation of land to a farmer till the harvest is sold. Farmers have to pay a license amount—license for working in the farms—to the panchayat and the government. The amount paid to the local panchayat is approximately Rs 1500/year and to the government Rs 50/bigha/year. Most of the times the panchayat members or their relatives end up receiving assistance and subsidies provided by the government. Further, culture farmers seen as belonging to or aligned to the ruling party are provided all kinds of assistance and security in carrying out fish farming activity.

Various government departments are also linked with the aquaculture activity in the region, though indirectly. For instance, the Department of Agriculture and the Department of Fisheries strive to ensure prosperous and sustainable agriculture, commercial fishing, aquaculture, and food production without causing harm to the environment. Their aim is to enhance economic stability, employment opportunities in coastal communities, and the sustainable use of coastal resources. The departments have to conserve, develop, and share the fish and other living aquatic resources of the region for the benefit of present and future generations. An increase in shrimp demand from processing/export units along with lucrative prices for the same have led to more and more land going into aquaculture. As a result, due to regular application of antibiotics, lime, and other organic, and inorganic fertilizers, the land is becoming infertile which is a serious cause of concern for the departments.

Likewise, the Forest Department has to ensure that mangroves are reserved for conservation purpose since they play a vital role in the coastal environment as a cyclone protection belt, as a habitat for juvenile fish and crustacean species, and through the supply of a variety of products (for example, shellfish, honey, and wood) to the local population.

Mangrove forests are one of the primary features of coastal ecosystems. Efforts made to utilize this resource have increased along with increasing demand, triggered by human population growth. Causes for the degradation of mangroves include land conversion, timber and fuelwood collection, grazing, and natural causes such as cyclone damage. With the increasing international demand for culture fish varieties, mangrove forest is converted into shrimp farms with communities settling down near these farms.

Further, MPEDA ensures that eco-friendly technologies are used in the aquaculture sector. It provides capital for the development of shrimp farms including farm construction, pumps, equipments, and machineries like aerators and generators. It also provides assistance for the purchase of water quality testing equipments.

Private Investors

Capital for aquaculture comes from diverse sources such as national banks, financial institutions, and the state government. With the increase in aquaculture practice in the Sundarbans, informal credit institutions like local moneylenders; aratdars, village panchayats, and relatives have come to play a significant role in providing capital for aquaculture purposes. These informal institutions are the preferred options farmers have been resorting to from time to time. Infact, the dependence of culture farmers on these institutions has grown over the years due to difficulties involved—cumbersome paper work—in receiving help from banks and other financial institutions. On the other hand, the gains from lending money to the farmers are quite substantial since the commission percentage on the amount advanced is high.

Post-Harvest Production Links

The off-farm post-harvest production links include processors and marketing agents (packaging, transport, exporters-importers, wholesalers and retailers, and restaurants).

Each commission agent is tied to a particular processing/export unit, which provides a loan in cash to him in return for procuring shrimp from individual fishers/farmers. The commission agent further lends this money to the farmers in return for the right to sell their products. Some traders, who do not borrow from processing/exporting firms, use their own funds for lending purposes, thus having an option of selling the produce to whichever unit is willing to pay them well. At times agents are also employed by the processing units to purchase raw material conforming to the price and quality specifications given to them by the unit.

LOCAL TRANSPORTERS

This is an important group, which has emerged due to an increase in the demand for aquaculture products by the exporting/processing units. Local transporters facilitate the movement of aquaculture produce from the local markets to the processing units. They ensure that the product is transported in a manner that prevents contamination and deterioration. Big processing units have their own refrigerated transportation vehicles while some depend on scooters and vans. With the frequent changes in the transportation requirements imposed by importing countries, EU in particular, the transporters have to invest huge amount in buying containers as per the specifications mentioned. For those firms who have their own transportation, complying with the changing requirements leads to inefficient use of their limited capital.

LOCAL MARKETS (DOMESTIC CONSUMPTION)

Given the domestic demand for aquaculture products, an aratdar might decide to sell some of the aquaculture produce in the local market. The demand generally comes from the local hotel industry. He may also decide to sell inferior quality shrimps at a lower price in the local market for domestic consumption. This normally happens in cases when he knows that the processing units will not accept the produce for exporting purposes. Nonetheless, he is able to find a market for the product. Moreover, shrimp being a luxury food item is very well taken in the local market even if it is of inferior quality.

National Government

In India, the Bureau of Indian Standards (BIS) has been designated as the WTO-technical Barriers to Trade Enquiry Point, while the Ministry of Commerce is responsible for implementing and administering the WTO agreement on Technical Barriers to Trade (TBT). The BIS, who endeavours to align Indian standards as far as possible with international standards, formulates Indian standards.

The Export Inspection Agency (EIA) is in overall charge of EU certification to be accorded to the processing units. The Export Inspection Council (EIC) acts as a facilitator and intermediary between the importing countries and export units. Therefore, any notification regarding Maximum Residual Limit (MRL), SPS, Product and Process Method (PPM), and Labelling and Packaging (L&P) received from importing countries is communicated to these units through EIC, which applies to all processing-cum-exporting units uniformly. It also advises the Government on measures to be taken to enforce quality control and inspection for exports. The EIA provide pre-shipment inspection and certification services.

MPEDA has to ensure that hygienic conditions prevail in the landing, pre-processing, and processing sectors. It is responsible for the promotion and regulation of exports of fish and fish products. It provides financial assistance for the following: acquisition of the fish processing machinery, installation of generating sets, establishment of chill room facility, installation of water purification system, and setting up of water effluent treatment plants.

Concluding Remarks

A set of national, state government, and departmental organizations provide infrastructural support to the export-oriented production and marketing of shrimp. The major actors, however, operate through well-functioning markets with a slew of agents and intermediaries linking exporting units with the hinterland from where aquaculturists provide shrimp, the main input into production. The processes of culture and collection in turn set in motion processes of change in land-use and in off-shore biodiversity. Income generation, land-use change and changes in off-shore biodiversity are three major consequences of this. The next few chapters examine the nature of these changes. In sum, we are trying to examine the impact of increasing export orientation on real people and real places, moving from macro-drivers towards more micro- and multi-dimensional impacts on human well-being.

NOTES

1. See 1st Working Plan, Dept. of Forests, Govt. of West Bengal, 1962.
2. The agricultural area is 2,95,000 ha. It is entirely mono-cropped due to the high salinity of the soil. Less than 1 per cent of the area is irrigated.
3. Data from MPEDA office in Kolkata.
4. The lower figure in the range comes from the survey of the processing units, which covered seven units producing together 65 per cent of total frozen shrimp production in the state. The higher figure is derived from the estimates of exports from Kolkata provided by MPEDA.
5. For details of data on land conversion, see Chapter 6 in which satellite data from 1986 to 2004 is analysed to discuss the drivers of land-use change.
6. A small hatchery was established by the West Bengal Fisheries Department in the coastal town of Digha in Midnapore district but it did not attain commercial proportions.
7. For more details on the kinds of gear used and location along creaks and rivers, see Danda (2007).

Appendix 4A

TABLES ON SOCIO-ECONOMIC CONDITIONS IN THE SUNDARBANS

Table 4.A1: Population by Different Categories of Work (Sundarbans and West Bengal)

Category	Sundarbans	West Bengal
	Million	Million
Number of households	0.69	15.87
Total population	3.76	80.18
Literate population (7 years and above)	2.03	47.20
Total workers	1.30	29.48
Main workers	0.91	23.02
Main cultivators	0.23	4.56
Main agricultural population	0.27	4.52
Main household industry workers	0.04	1.44
Other main workers	0.37	12.51
Marginal workers	0.39	6.46
Marginal cultivators	0.08	1.10
Marginal agricultural population	0.20	2.84
Marginal household industry workers	0.03	0.74
Marginal other population	0.09	1.78
Non-working population	2.45	50.69
Total agricultural workers	0.47	7.36

Source: Directorate of Census Operations (2001).
Note: Total workforce= main workers + marginal workers
Total population= total working population + non-working population
Total agricultural workers= main agricultural workers + marginal agricultural workers

Table 4.A2: Availability of Sanitation and Electricity (West Bengal and Sundarbans)

Area Name		Total Number of Households	Total Population	Electricity Available to Number of Households	Electricity Available to Population (per 1000)	Toilet Facility Available to Number of Households	Toilet Facility Available to Population (per 1000)
West Bengal	Total	15,715,915	1,000	58,85,724	370	68,69,777	435
				37%	37%	44%	44%
Sundarbans	Total	6,86,215	19,000	42,940	1,096	2,02,236	5,701
				6%	6%	29%	30%

Source: Directorate of Census Operations (2001).

Table 4.A3: Sources of Drinking Water (West Bengal and Sundarbans)

Area Name	Number of Households	Tap	Handpump	Tubewell	Well	Tank, Pond, Lake	River, Canal	Spring	Any Other
West Bengal	1,57,15,915	33,63,802	87,71,743	17,77,159	15,68,840	34,236	42,302	1,02,296	55,537
		21.4%	55.8%	11.3%	10.0%	0.2%	0.3%	0.7%	0.4%
Sundarbans	6,86,215	45,203	5,50,344	85,527	109	3,012	157	1,547	316
		6.59%	80.20%	12.46%	0.02%	0.44%	0.02%	0.23%	0.05%

Source: Directorate of Census Operations (2001).

5

BIODIVERSITY LOSS OFF THE SUNDARBANS COAST
MAGNITUDE, COST, AND IMPACT

INTRODUCTION

Shrimp farming and aquacultural activities are carried out mostly in coastal regions as they provide vast tracts of saline land coupled with abundant quantity of wild seeds. National governments are supporting this activity in the belief that shrimp farming can generate significant foreign exchange earnings, and enhance employment opportunities and incomes in poor costal communities. As a consequence, hundreds of hectares of land has been brought under this venture. But this expansion has several effects on the land and water regimes and is postulated to lead to the degradation of the marine environment. Biodiversity, for instance, is impacted by the practice of catching post-larvae shrimp, which has detrimental effects for other species.

Increase in the practice of aquaculture in the Sundarbans region during the last decade or more together with the absence of hatcheries resulted in the seed input (seed of tiger prawn) being collected from the wild, using labour–intensive drag-nets of different kinds. It has been reported in several studies that during the collection of tiger prawn seeds, juveniles of many species of finfish and shellfish are trapped in the net and these non-target species are thrown away and destroyed, as they are not remunerative. This destructive practice causes a major damage to juvenile finfish community of the area. The juvenile stage of finfish community referred to as *icthyoplankton* constitutes an important planktonic component of the marine and estuarine ecosystems and forms an integral part of the fresh-cum-brackish-cum-saline water owing to their migratory behaviour. These planktons in a sense are referred to as *ecological crop* of the marine and estuarine systems as they provide nutrition to members of higher trophic level, which includes larger bony fishes, sharks, turtle, dolphin etc.

The demand for tiger prawn seeds has risen exponentially and, in the absence of any hatchery, the entrepreneurs have no option but to depend on wild harvest of tiger prawn seeds, carried out by rural prawn collecting households. Several studies purport to estimate this loss. It is reported, for instance that about 48 species of finfish juveniles are wasted per net per day per haul, which amounts to about 9.834 kg.[1] This constitutes a huge loss of species diversity.

Further, the sustainability of shrimp farming is also threatened by its reliance on the collection of wild shrimp fry. This activity sustains a large number of poor households, using cheap and destructive methods to supply the key seed inputs to shrimp farmers, but these methods may, in the process, be damaging wild stocks of both shrimp and other aquatic species. Although hatcheries are being developed as a potential alternative to the supply of seed from the wild, their development has not been as rapid as desired resulting in growing dependence on wild seed stock. The Sundarbans region in West Bengal is experiencing loss in fish diversity due to excessive collection of shrimp fry. Destruction of aquatic resources is considerable due to harmful practices in the discard of by-catch.

The questions then arise:

- Can this social cost of biodiversity loss, due to prawn seed collection, be internalized into the private cost of a shrimp farmer given the present cost structure under which he operates?
- How significant is the input which could be called 'biodiversity use' in the total cost of production? How would it affect his per unit cost and hence profit?
- Do the substitution possibilities between inputs get affected by such internalization?
- Do scale economies exist in the aquaculture production and how does inclusion of biodiversity cost affect their magnitude?

This chapter examines these issues. A methodology based on the translog cost function is used for the purpose. Section on 'Indices of Ecological Crop Loss' in this chapter reviews time-series estimates of the ecological crop loss and estimates trends in biodiversity loss based on time-series indices for representative sites. The next section examines different approaches to internalization of the biodiversity loss caused by seed collection technology and uses a cost function approach to estimate it. The database is described and results analysed in the subsequent two sections. Finally, the last section provides pointers towards policy options.

INDICES OF ECOLOGICAL CROP LOSS

Two kinds of factors have operated in the period since 1991 to change the magnitude of loss of biodiversity, otherwise referred to as loss of by-catch or ecological crop loss. They are:

1. The steady increase in land under aquaculture would be expected to have increased the loss. About 33,000 hectares in North 24 Paraganas and 12,000 hectares in South 24 Paraganas are devoted to shrimp farming. Potential areas which could be brought under aquaculture is estimated at 1,80,000 hectares in these two districts.[2] Future expansion could accentuate this ecological crop loss substantially.
2. Simultaneously however, the substitution of improved traditional system of shrimp culture for semi-intensive culture following the Supreme Court of India order of 1996 has reduced stocking density per hectare and consequently the loss in by-catch or ecological crop loss per hectare.

While one-time estimates of 'ecological crop loss exist, it is important to understand changes over time and space underlying this loss. In other words, what was the extent of change in biodiversity indices in the 'by- catch' from tiger prawn seed collection during the decade of the nineties?

The expansion of aquaculture in the Sundarbans has been accompanied by fast market integration and the rapid movement of raw materials and output across the region. Transport of both raw material and output is well-organized and efficient.[3] We assume therefore that seed collection is spread over the entire region. A representative selection of sites for estimation of biodiversity loss therefore needs to take into account varying salinity levels and dilution factors which vary in different locations of the coastal zone.

Monthly data for ten years on number of species lost in the by-catch from three representative sampling stations was used to estimate the indices of biodiversity loss.[4] The sample size for the computation was a 10gram composite sample of the wasted material obtained by random mixing of the collection of 15 nets. The three sampling stations were selected for their different salinity profiles and distinct identity. These are: Diamond Harbour (Station 1), Sagar South (Station 2), and Junput (Station 3). We give below the characteristics of these three stations.

Diamond Harbour is situated in the low saline upper stretch of Hugli estuary, just outside the northern boundary of the Indian Sundarbans. The station is very near to the Haldia port-cum industrial complex. Salinity of surface water is minimal around the station due to its location away from the Bay of Bengal in the extreme upstream region and also due to huge water discharge from the Hugli river, which is perennial in nature. The station has no mangrove vegetation except for a few mangrove associates and seaweeds.

Sagar South is situated on the south-western tip of Sagar island and falls in the western sector of the Indian Sundarbans. The station has rich mangrove vegetation and extensive mud flats. Although there is no industrial activity around this station, the presence of a large number of shrimp culture farms (carrying on traditional culture with low stocking density) has enriched the surrounding water with nutrient and organic load.

Junput is situated in the Medinipur district of coastal West Bengal and is noted for its high aquatic salinity due to its proximity to the Bay of Bengal. The extremely high salinity has posed an inhibitory effect on the growth and survival of mangroves in the region. Existence of salt pans in the vicinity has made the soil of this region hyper-saline in nature. The presence of Digha Tourist centre and Shakarpur fishing harbour close to the station is the source of anthropogenic pressure, though there is no industry in the vicinity.

The ecological crop loss was assessed in terms of three alternative indices:

The Shannon Weaver species diversity index: $H = \Sigma n \Sigma ni/N \ln ni/N$

Where ni = total of importance value of each species (number of individuals of each species),

N= total of importance values, that is, total number of individuals of all species in the wasted sample.

Index of dominance is given by $(\Sigma (ni/N)**2)$

Evenness index is given by H/lnS, where S is the number of species.

The results are tabulated in Tables1 to 3 in Appendix 5A. The 10-year monthly data reveals seasonal oscillation. The values of diversity are highest during the pre-monsoon period (March to June) and lowest during the monsoon (July to October). The seasonal trend is due to the lifecycle of most organisms in the ecosystem.

Further, linear regressions using biodiversity index as the dependent variable on time were run for the three sampling stations namely, Diamond Harbour, Sagar South, and Junput. For the two stations, Diamond Harbour and Junput, the trend coefficients came out to be negative indicating that there is an increasing trend in biodiversity loss due to prawn seed collection. This is indeed an important result. In one case, the coefficient did not indicate a significant negative trend. This was in Sagar South where there exists extensive mangrove vegetation and very little anthropogenic pressure. It is known that mangroves are nurseries, both for shrimp seed and also harbour large number of species.

The results from the analysis of the ten-year data can be interpreted thus: biodiversity loss due to prawn seed collection is likely to be far more in regions where it takes place together with other anthropogenic pressure, resulting in land conversion away from mangroves. Since rapid transport of prawn seed takes place all over the region, allocating the loss spatially is difficult. The average annual decline 0.03% was taken to arrive at the figure of biodiversity loss over time. To allocate this decline to individual farms, it was assumed that a particular farm's contribution to biodiversity loss was proportional to its demand for seed, which depends on its stocking density and its size. The higher the stocking density, the more is its demand. Similarly, even with the same stocking density, larger farms make a larger dent on loss of biodiversity.

INTERNALIZING THE COST OF BIODIVERSITY LOSS USING TRANSLOG COST FUNCTION

The Issues and the Methodology

The erosion of biodiversity as a consequence of the mode of seed collection is a social cost of the 'technology' adopted for provision of one of the critical inputs. This erosion of the quality of aquatic resources is an unintended environmental impact of economic activity, often referred to as 'an externality'. An externality exists whenever one individual's actions affect the well-being of another individual or of the environment—whether for the better or for the worse—in ways that need not be paid for according to the existing definition of property rights and other related legal regimes prevalent in the society. To elaborate, an 'external diseconomy', 'external cost' or 'negative externality' results when part of the cost of producing a good or service is borne by a firm or household or a component of society other than the producer or purchaser. The literature of environmental economics throws up a number of cost-related notions to refer to and to understand these unintended negative consequences.

The definition of what constitutes private cost to a firm depends on the legal framework within which it operates. Since the damage done to the environment is outside the reckoning, understanding or calculus of the firm or the farm, it is not part of its cost, generally referred to as 'private cost'. These costs are typically the cost of inputs such as labour, land, and raw material which it has to incur. However, a change in the legal environment within which the firm is operating can result in a change in this. If the marketing of a firm's product is contingent on the use of a set of processes or technologies due to the existence of 'environmental standards', the firm's private cost of production may be increased by what are often referred to as 'costs of compliance' (with legal standards). The existence of international safety and SPS standards often results in such costs for processing industry in the context of shrimp export. These compliance costs are inevitably a part of firms' private costs of production, provided of course that the standards are implemented without exception.

Costs of off-shore biodiversity loss caused by collection of PL from the wild are not of the nature of compliance costs under the present legal dispensation. If however, eco-labelling sets a standard such that the export of shrimp produced from seed collected from the wild becomes illegal, this social cost would be automatically internalized and become compliance cost for the shrimp farmer. In other words, what counts as private costs to the producer and gets accounted as such or 'internalized' depends on the product standards in prevalence, both for output and input: in other words, on the legal regime within which the producer is operating.

The question being asked in this exercise is different, though related. It deals with the economic feasibility of internalizing the cost of 'biodiversity or ecological crop loss' associated with the current technology of shrimp farming, in which shrimp seed used is collected from the wild. We discuss here whether the structure of the cost function and the parameters within which the shrimp farmer operates would permit him to internalize this social cost in his private reckoning as an increased cost of seed. Would the shrimp farm still be a 'going concern' if he had to internalize the cost of off-shore biodiversity loss by paying more for it?

To elaborate, the institutional structure through which the social cost is measured and passed on to the shrimp farmer is not discussed. It could be measured, for example, in one of the following ways:

1. The purchase price of seed could go up if off-shore collection were to be banned and hatchery-produced seed alone were available,
2. The purchase price may go up if higher capital cost was incurred by PL collectors in purchase of nets of better quality which eliminated by-catch of non- prawn species, and
3. Prevention of social cost could also occur if better water treatment (accompanied with higher labour use and cost) were to reduce the demand for seed quantity per unit of output of shrimp, thereby reducing seed extraction from the sea and hence biodiversity loss.

Other routes for operationalzing internalization cost could also be figured out as better social arrangements emerge. The present exercise addresses the following questions:

1. How significant is the biodiversity use input in the total cost of production? How would it affect his per unit cost and hence profit?
2. Do the substitution possibilities between inputs get affected by such internalization?
3. Do scale economies exist in the aquaculture production and how does inclusion of biodiversity cost affect their magnitude?

In this section, we address the above questions by using a cost function framework. Such a framework for examining the relationship between environmental and other production costs has been used recently by Morgenstern, Pizer, and Shih (2001). The analysis follows earlier established methods for studying substitution between natural resources, capital, and labour[5] using cost and production functions.

The cost function can be defined as the function specifying the minimum costs of producing an output with a given vector of input prices and a technology. The duality relation between the cost function and the production technology is used to specify the cost function. A translog cost function has been estimated, as it allows scale economies to vary with the level of output and also it does not impose restrictions on substitution possibilities between the factors of production. A recent application in the Indian context to examine the hidden costs of environmental regulation in the textile industry is by Tholkappian (2005).[6]

The following form of translog cost function is chosen for estimation in the present study:

Ln TC=

$$\beta_0 + \beta_y \ln Y + \sum_i \beta_i \ln P_i + \frac{1}{2}\beta_{yy}(\ln Y)(\ln Y) + \frac{1}{2}\sum_i \beta_{ii}(\ln P_i)(\ln P_i) + \sum_i \sum_j \beta_{ij}(\ln P_i)(\ln P_j) + \sum_i \beta_{iy} \ln P_i \ln Y$$

Where, $\beta_{ij} = \beta_{ji}$, TC= total cost, β_0 = constant term, Y=Output, P_i = vector of input prices.

In order to improve the efficiency of the estimates, translog total cost function is estimated along with share equations. Differentiating the total cost function with respect to input prices can arrive at the share equations for each factor. The resulting share equation (S_i) takes the following form:

$$\frac{\delta \ln TC}{\delta \ln Pi} = \beta i + \beta ii \ln Pi + \sum_i \beta ij \ln Pj + \beta iy \ln Y$$

The specified cost function and the share equations are estimated jointly, applying the non-linear maximum likelihood method. To overcome the problem of singularity, one of the share equations (waterfeed equation in present case) is arbitrarily dropped from the system estimation. The resulting maximum likelihood estimates are invariant to the equation deleted.

From translog cost function, the Allen partial elasticities of substitution (σij) and (σii) for the ith factor of production are calculated as:

$$\sigma ij = \frac{(\beta ij + Si\, Sj)}{(Si Sj)}, \qquad i \neq j$$

$$\sigma ii = \frac{\beta ii + (Si^*Si) - Si}{Si^*Si}$$

Price Elasticities of demand for factors of production:

Own
$Eii = Si\, \sigma ii$

Cross
$Eij = Sj\, \sigma ij$

THE DATABASE

The data used is derived from the primary survey conducted by the project team during February 2005. Aquaculture farms located in three blocks of the Sundarbans deemed to be representative of the varying conditions under which aquaculture is carried out in the region.

On the basis of secondary data on water pollution[7] and information gathered from various sources regarding the shrimp business in the Sundarbans, following areas have been identified for the survey to study the link between shrimp production and water pollution:

Areas were selected on the basis of water pollution parameters exceeding the standards and presence of large area under shrimp farming.

1. Minakhan: The analysis of data shows that Minakhan has relatively higher compound growth rate (CGR) for DO (Dissolved Oxygen) than other stations but has a low average DO (International Standard of DO for aquatic life is 5 mg/l (minimum)) for many periods in the 1999–2003 period, which is also substantiated by studies done on Minakhan estuary. In this area, CGRs for turbidity and nitrate is also highest along with highest average turbidity for all the periods in 1999–2003. About 3,600 ha of area is under shrimp production.
2. Canning: The area shows relatively higher CGRs for DO and turbidity than other stations for the period 1999–2003. The average levels of DO are marginally below the standard in some years. There are around 150 ponds for shrimp farming at Canning.

Fifty aquaculture farms from three blocks, Canning, Minakhan, and Gosaba were surveyed. Out of these two, farms are dropped out for the analysis since they are not culture ponds. For the remaining 48 farms, data was collected on different aspects of production, input costs, and technology of production for the year 2004.

Inputs for which use and cost data was collected are: feed, chemicals/fertilizers used in water treatment, land, and seed. In addition, there is information on stocking density, size of the farm, lease rate, labour employed (permanent and temporary) on the farm, their wage rates, shrimp production, selling price etc.

For the cost function estimation, the inputs used are: water and feed, seed, land, and labour. The cost function and share equations are estimated with a term for value of production for each farm in order to examine scale economies.

Additionally, the estimation is carried out taking into account the additional cost of biodiversity loss consequent on the mode of seed collection. Details with respect to the data collected are given below.

Water-feed Cost: Price per Hectare of Chemicals/Fertilizers Used in Water Treatment, Feed Used and Cost per Farm

A large majority of the farms use lime, bleaching powder, ammonia, and urea for treating the water before stocking juveniles. These fertilizers not only improve the quality of water but they also act as feed for shrimp PL. At times, some of the farmers do buy artificial feed from the market over and above this. Even then it was easier to combine the two inputs since the quantity (ha) and price (Rs/ha) were given in the same unit. The corresponding cost of waterfeed is computed by multiplying price (Rs/ha) by pond size.

Rent per Hectare and Cost of Land per Farm

The farms are taken on lease, mostly for a three-year lease period, with lease amount varying for each farm depending on its location. A farm closer to the water source has to pay a larger amount as lease. The data reveals a large variation in the lease rate. The rate per hectare varies from Rs 9,231 to Rs 76,923 per year. The corresponding cost of land is computed by multiplying lease rate by the area of the pond.

Wage Rate and the Cost of Permanent Labour Employed

The number of family members working on the farm and the number of hired permanent labour (works throughout the year) is recorded separately and the payments to the hired labour are given. A wage rate per person for permanent labour is worked out for each farm and family labour is also valued at this wage rate to yield cost of labour in aquaculture production.

Stocking Density and the Cost of Seed per Farm

Stocking density varies from farm to farm depending on the culture system (technique) adopted. At present, the farms mostly follow extensive and improved traditional systems. Stocking density is low, little management is done through periodical water exchange during high tide, generally twice a month. The farm is fertilized at low dose rates. Sometimes supplementary feed is used. Seed price varies from Rs 210 per thousand to Rs 1,200 per thousand. Seed cost per farm is estimated by multiplying price by the seed stocked in thousands.

COST OF 'BIODIVERSITY EROSION' (TREATED AS AN INPUT) PER FARM

The biodiversity loss cost is estimated using the ecological crop loss assessed in terms of diversity (Shannon Weiner species diversity index, H) to evaluate the quantum of damage both in terms of weight and variability. The average trend decrease in this index is a measure of 'biodiversity erosion' in the region. It is attributed to individual farms in proportion to their stocking density and farm size. In our analysis, seed cost and 'biodiversity erosion' cost have been combined to arrive at one cost. It could easily be done since both depend on stocking density. Also, the seed price is given in Rs/000 similar to what we have estimated for 'biodiversity erosion' cost. In order to arrive at the cost per farm of the 'biodiversity erosion', shrimp production per farm is multiplied by the figure 0.03 representing trend decrease in biodiversity index. This would give us the cost per farm of 'biodiversity eroded' in terms of rupees. For a translog cost function price of the 'biodiversity erosion' is required. Using the following formula, the same has been arrived at:

$$\text{BD Price (Rs/000)} = \left(\frac{\textit{Value of shrimp production}}{\textit{stocking density}} \right) * 0.03$$

Average cost shares for the four inputs are given in Table 5.1. Labour, land, and seed-biodiversity inputs comprise 35, 32 and 27 per cent of the total cost of production per hectare. Water management and feed together contribute only about 5.8 per cent.

Table 5.1: Cost Shares of Inputs in Aquaculture

Inputs	%
Water-feed	0.0583
Labour	0.3512
Landlease	0.3224
Biodiversity erosion-seed	0.2681

Source: Authors' calculations.

RESULTS AND ANALYSIS

In the present model, the social cost of biodiversity loss arising out of the mode of seed collection is internalized in the total cost of aquaculture by including it in the cost of seed collection.

MODEL WITH BIODIVERSITY COST INTERNALIZED IN THE INPUT COST FOR SEED

The translog cost function and associated system equations for the model with biodiversity cost internalized are given in Tables 5.2 to 5.7. We use the following notation:

Wf— Water and feed prices combined together (Rs/ha)
La— Wages to permanent Labour+ family members working as labourers (Rs/person)
Le— Land lease amount (Rs/ha/year)
Bd— Biodiversity erosion cost and seed prices combined together (Rs/000)
Y— Value of output (Rs)

Table 5.2: Estimates of Translog Total Cost Function

Variable	Coefficient	T-ratio
A	15.789	1.994
βla	1.039	3.978*
βle	−0.495	−2.522**
βbd	0.405	1.280
βy	−1.207	−0.972
$\beta lala$	0.094	2.449**
$\beta lale$	−0.034	−4.219*
$\beta labd$	−0.054	−1.434
$\beta lele$	0.042	6.431*
$\beta lebd$	0.001	0.061
$\beta bdbd$	0.055	1.347
βlay	−0.060	−2.905*
βley	0.070	4.450*
βbdy	−0.011	−0.432
βyy	0.149	1.542

Source: Authors' calculations.
Note: *, **, and *** denote significance at 1, 5, and 10 per cent.

The estimated cost function is a well-behaved cost function, as the fitted factor shares are positive at almost all the observations. Labour has the highest share in the total cost followed by land lease, biodiversity erosion-seed, and water-feed, in that order.

Table 5.3: Own Elasticity of Substitution (Allen)

Variable	Coefficient
Water-feed	−11.28
Labour Labour	−1.086
Landlease Landlease	−1.69
Biodiversity erosion-seed	−1.95

Source: Authors' calculations.

The own elasticities for all the inputs, that is, waterfeed, labour, lease, and biodiversity are negative (Table 5.3). However, while labour (wage rate) and lease (lease rate) are significant determinants of cost, price of seed and 'biodiversity erosion' costs are not coming out to be significant at 5 per cent and 10 per cent as can be seen from Table 5.4.

Table 5.4: Own Price Elasticity of Demand

Variable	Coefficient
Water-feed	−0.65
Labour Labour	−0.38
Landlease Landlease	−0.54
Biodiversity erosion-seed	−0.52

Source: Authors' calculations.

Own price elasticities are negative for all the inputs. None of the inputs have high price elasticities. This implies that increases in the prices of the inputs shall not impact their demand much. Water-feed having the highest elasticity is not very responsive to its own price- own price elasticity being −0.65. Next come land and seed-biodiversity with own price elasticities in the range of −0.52 to −0.54. On the other hand, labour has the lowest own price elasticity (−0.38) indicating that the responsiveness of labour demand to its own price is very low.

These findings are important with respect to the social cost of biodiversity loss and its delegation to aquaculture farmers. *These results indicate that even if a larger biodiversity cost were to be assigned to aquaculture and seed prices were to rise as a consequence, the farmers*

Table 5.5: Cross-price Elasticities

Variable	Coefficient
Water-feed Labour	0.24
Labour Water-feed	0.04
Water-feed Landlease	0.17
Landlease Water-feed	0.03
Water-feed Biodiversity erosion-seed	0.23
Biodiversity erosion-seed Water-feed	0.05
Labour Landlease	0.22
Landlease Labour	0.24
Labour Biodiversity erosion-seed	0.11
Biodiversity erosion-seed Labour	0.14
Landlease Biodiversity erosion-seed	0.27
Biodiversity erosion- seed Landlease	0.32

Source: Authors' calculations.

would be able to absorb it, given the present structure of costs and the present price levels for their output.

Overall cross-price elasticities are low, meaning demand for inputs is inelastic to change in the price of other inputs (Table 5.5). However, some conclusions can be drawn. These are:

Water-feed is responsive, though not very strongly, to the change in the price of labour as can be seen from the above table, water-feed labour—0.246. Labour can be substituted, to a limited extent by water management and feed.

Cross-price elasticities between water-feed and lease are low. It would be difficult to draw any conclusion. Though one can say that as the lease rate increases, demand for more land under aquaculture production goes down. Without increasing the size of the cultured area, same output can be produced by improving water quality and increasing the quantity of feed. The relationship is not very strong since the elasticities are low (0.172 and 0.03).

Cross-price elasticities between lease (price of land) and biodiversity (price of seed and 'biodiversity erosion') show that the change in the price of one leads to change in the demand for other. Though here again the elasticities are low.

Table 5.6: Cross Elasticity of Substitution (Allen)

Variable	Coefficient
Water-feed Labour	0.70
Water-feed Landlease	0.53
Water-feed Biodiversity erosion-seed	0.89
Labour Landlease	0.70
Labour Biodiversity erosion-seed	0.42
Landlease Biodiversity erosion-seed	1.01

Source: Authors' calculations

Table 5.6 gives the elasticities of substitution between inputs used in aquaculture. A positive value indicates that the inputs are substitutes, a negative then the two inputs are complementary with each other. The production structure reveals interesting possibilities of substitution.

Land lease and biodiversity are strongly substitutable with AES (Allen elasticity of Substitution) equal to 1.007. This only states the fact that if land were to be in short supply, output can be increased by increasing the stocking density. They are substitutable. This would have implications for biodiversity loss. Conversely, if you use more land and extensive or improved traditional technologies, lower stocking densities and hence less loss of biodiversity is implied. *A land-intensive aquaculture expansion is indicated if biodiversity loss is to be averted.*

However, since the cross price elasticities between the price of land and of seed- biodiversity are not high, this effect is likely to be limited in magnitude.

The same is true for water management and feed input (considered as one input in this model) Water-feed and seed-biodiversity are substitutable—the estimated AES is 0.89. However, the cross price elasticities Water-feed and seed-biodiversity and seed-biodiversity and water-feed are very low at about 0.238 and 0.051.

Water-feed and Labour display some substitutability—AES is 0.701 and cross price elasticities, 0.24 and 0.04. Looking at the cross price elasticities, one can say that limited substitutability exists if price is to be its driver.

Labour and land lease are also found to be substitutes—AES is 0.701 and cross price elasticities 0.226 and 0.246 respectively.

ECONOMIES OF SCALE

The elasticity of total cost, which is a proportional increase in total cost (TC) resulting from a small proportional increase in output (Y), is calculated by differentiating the total cost function with respect to output.

Scale economies (SCE) is defined as:

$$SCE = \frac{\delta \ln TC}{\delta \ln Y}$$

Further, the negative coefficients of βlay and βbdy imply that with the increase in the scale, less of labour is required (Table 5.7). Similarly, as the scale of operation increases, it leads to less than proportionate increase in seed and biodiversity use, which in turn means less of biodiversity loss. This is a significant finding, implying as it does that if the current technology of aquaculture production were to be used on larger scales, it would provide the advantage of economies of scale without resulting in proportionate loss in biodiversity.

Table 5.7: Estimates of Translog Total Cost Function

Variable	Coefficient
βlay	–0.06
βley	0.07
βbdy	–0.01
βwfy	0.001

Source: Authors' calculations.

There is evidence of strong scale economies in the aquaculture production. The value of 0.54 of the coefficient of economies of scale indicates that a 1 per cent increase in the output would lead to less than 1 per cent increase in the total cost. Larger aquaculture farms are therefore indicated.

When this characteristic of the production technology is interpreted together with the finding that land-intensive aquaculture reduces biodiversity loss, setting up of large aquaculture farms is indicated as the policy direction.

CONCLUDING REMARKS AND POLICY IMPLICATIONS

Biodiversity loss can be mitigated through innovative policies. The following policy options exist and may be considered:

- Internalize the cost of biodiversity loss in the aquaculture farming cost. This chapter illustrates that if more land is used and extensive or improved traditional technologies adopted, lower stocking densities and hence less loss of biodiversity is implied. *A land- intensive aquaculture expansion is indicated if biodiversity loss is to be averte*d. Further, the existence of economies of scale

in aquaculture production points towards the economic viability of such an approach.
- The above is best implemented by setting up a manner of collecting additional charges on aquaculturists in a 'designated local fund' to be spent on biodiversity conservation. This is economically feasible but administratively difficult to implement.
- A market-administered policy option is to aim at eco-labelling for the acceptance of processed shrimp both for export and for domestic consumption. The label would require, among other things, that the prawn seed be sustainably harvested. Given the cost structure found to exist, this is eminently feasible, in particular for export markets.
- A possible option is to legally ban 'seed collection from the wild' and provide prawn seed through the 'hatcheries technology'. This would need to be supplemented by providing alternative livelihoods to prawn seed collectors. Such provision of alternative livelihoods is important because prawn seed collection is an important livelihood for majority of the poor. Such an option also requires extensive preparation by way of setting up and enabling social and legal frameworks.

NOTES

1. See Mitra (2005), MOEF, and Project Report (1996).
2. Government of West Bengal (2004).
3. This is well-documented. The research team also came across evidence of this efficient transport network during the course of its visits to the region.
4. This rich data set was made available and analysed due to the collaboration of the Department of Marine Science, University of Kolkata. See Mitra (2005).
5. See the papers by Humprey and Moroney (1975) and Berndt and Wood (1975).
6. For earlier discussions on methodological issues in the context of water pollution in India see Goldar, Misra and Mukerji (2001). An early application of the Translog cost function to Indian agriculture is in Chopra (1985).
7. Provided by Abhijit Mitra, Department of Marine Sciences, Calcutta University.

APPENDIX 5A

TABLES-SPECIES DIVERSITY, EVENNESS, AND DOMINANCE

Table 5A.1: Species Diversity Indices (Diamond Harbour)

Year	Jan–March	April–June	July–Sep	Oct–Dec
1994	8.4308	8.7809	6.3189	8.4474
1995	8.4143	8.8054	6.3090	8.4644
1996	8.4283	8.7134	6.2239	8.3684
1997	8.2988	8.6932	6.1192	8.3294
1998	8.5144	8.8117	6.4153	8.5394
1999	8.5143	8.8155	6.4119	8.5052
2000	8.4793	8.7904	6.3472	8.4607
2001	8.4926	8.8137	6.3472	8.4773
2002	8.4793	8.8357	6.3472	8.4607
2003	6.6878	8.9534	0.0000	5.1895

Source: Mitra, 2005.

Table 5A.2: Species Diversity Indices (Sagar South)

Year	Jan–March	April–June	July–Sep	Oct–Dec
1994	9.7019	10.7737	4.7281	7.9539
1995	9.6321	10.7768	4.0187	7.9488
1996	9.6412	10.7825	4.2385	8.0033
1997	9.6717	10.6998	4.1653	8.0504
1998	9.6430	10.7726	3.8308	7.9658
1999	9.6559	10.7646	4.2748	8.0727
2000	9.7067	10.7897	4.3052	8.0588
2001	9.6767	10.8059	4.1698	8.0540
2002	9.7888	10.2239	7.0940	9.1811
2003	9.6675	10.7795	4.2015	8.0339

Source: Mitra, 2005.

Table 5A.3: Species Diversity Indices (Junput)

Year	Jan–March	April–June	July–Sep	Oct–Dec
1994	10.0974	10.4838	6.2160	9.3385
1995	10.0715	10.4803	6.2032	9.3573
1996	10.0941	10.4877	6.2360	9.3884
1997	10.1032	10.4720	6.2255	9.4084
1998	10.1046	10.4725	6.2497	9.4111
1999	10.1390	10.3894	6.1920	9.4153
2000	9.9996	10.4266	6.2433	9.3655
2001	10.1497	10.4719	6.2693	9.4276
2002	10.1496	10.5223	5.9456	9.4256
2003	9.6326	10.4214	4.7341	8.6465

Source: Mitra, 2005.

Table 5A.4: Indices of Evenness (Diamond Harbour)

Year	Jan–March	April–June	July–Sep	Oct–Nov
1994	2.6536	2.7170	2.5952	2.6998
1995	2.6484	2.7131	2.5906	2.7048
1996	2.6527	2.6959	2.5541	2.6745
1997	2.6141	2.6899	2.5124	2.6611
1998	2.6799	2.7265	2.6358	2.7147
1999	2.6799	2.7277	2.6345	2.7186
2000	2.6688	2.7199	2.6076	2.7042
2001	2.6731	2.7271	2.6076	2.7094
2002	2.6688	2.7224	2.6076	2.7042
2003	2.8270	2.8176	0.0000	2.7092

Source: Mitra, 2005.

Table 5A.5: Indices of Evenness (Sagar South)

Year	Jan–March	April–June	July–Sep	Oct–Nov
1994	2.7746	2.8492	2.4245	2.7097
1995	2.7713	2.8500	2.2320	2.6883
1996	2.7739	2.8515	2.3543	2.6987
1997	2.7827	2.8298	2.2772	2.7134
1998	2.7743	2.8490	2.1588	2.7117
1999	2.7859	2.8468	2.3710	2.7197
2000	2.7928	2.8535	2.3838	2.7267
2001	2.7839	2.8577	2.3801	2.7142
2002	2.8019	2.7858	2.6114	2.8004
2003	2.7815	2.8507	2.3125	2.7080

Source: Mitra, 2005.

Table 5A.6: Indices of Evenness (Junput)

Year	Jan–March	April–June	July–Sep	Oct–Nov
1994	2.8108	2.8415	2.7660	2.8056
1995	2.8036	2.8405	2.7607	2.7927
1996	2.8099	2.8426	2.7728	2.8020
1997	2.8126	2.8382	2.7689	2.8084
1998	2.8129	2.8384	2.7808	2.8090
1999	2.8150	2.8357	2.7561	2.8104
2000	2.8132	2.8455	2.7768	2.8261
2001	2.8179	2.8446	2.8072	2.8138
2002	2.8253	2.8520	2.6310	2.8467
2003	2.8510	2.9002	2.6445	2.7286

Source: Mitra, 2005.

Table 5A.7: Indices of Dominance (Diamond Harbour)

Year	Jan–March	April–June	July–Sep	Oct–Nov
1994	0.2238	0.2012	0.4623	0.2317
1995	0.2228	0.1999	0.4637	0.2305
1996	0.2235	0.2083	0.4898	0.2425
1997	0.2316	0.2089	0.5041	0.2460
1998	0.2146	0.1989	0.4367	0.2209
1999	0.2140	0.1980	0.4366	0.2230
2000	0.2174	0.2002	0.4528	0.2286
2001	0.2158	0.1972	0.4528	0.2264
2002	0.2174	0.1974	0.4528	0.2286
2003	0.3638	0.1786	1.0000	0.6012

Source: Mitra, 2005.

Table 5A.8: Indices of Dominance (Sagar South)

Year	Jan–March	April–June	July–Sep	Oct–Nov
1994	0.1438	0.0938	0.8193	0.2658
1995	0.1470	0.0932	1.0589	0.2740
1996	0.1464	0.0930	0.9932	0.2693
1997	0.1455	0.0981	1.0293	0.2654
1998	0.1473	0.0937	1.1701	0.2731
1999	0.1464	0.0940	0.9853	0.2626
2000	0.1414	0.0933	0.9610	0.2585
2001	0.1437	0.0922	0.9901	0.2645
2002	0.1325	0.1131	0.3844	0.1648
2003	0.1451	0.0931	1.0160	0.2668

Source: Mitra, 2005.

Table 5A.9: Indices of Dominance (Junput)

Year	Jan–March	April–June	July–Sep	Oct–Nov
1994	0.1189	0.1021	0.4630	0.1573
1995	0.1203	0.1025	0.4656	0.1576
1996	0.1193	0.1021	0.4639	0.1562
1997	0.1194	0.1031	0.4637	0.1550
1998	0.1195	0.1031	0.4594	0.1550
1999	0.1182	0.1064	0.4721	0.1556
2000	0.1227	0.1043	0.4586	0.1542
2001	0.1169	0.1025	0.4524	0.1521
2002	0.1165	0.1003	0.5370	0.1511
2003	0.1353	0.1011	0.7849	0.2106

Source: Mitra, 2005.

6

Land-use Change in the Sundarbans

LAND-USE CHANGE

Land-use is a very meaningful indicator for the characterization of any ecosystem and holds special relevance for a fragile ecosystem like mangroves. Land-use types can be characterized further through other indicators with respect to the ecological inventory. The formulation of land-use types can give an overview of the ecological situation in the region. With growing trade and direct investment pouring into activities like aquaculture, an increase in 'ecological footprint', or the area of productive land needed to sustain a defined population indefinitely is indicated. The growth of activities like aquaculture induces incentives for the farmers and other landowners in the adjacent areas to convert (sell) their land for shrimp farming since the returns from agriculture or other uses proves to be far lower than the returns from shrimp farming.

Anthropogenic pressures on mangrove ecosystems have caused a number of undesirable changes in the health and resilience of those ecosystems sustaining people's livelihoods. Problems like biodiversity loss, eutrophication, acidification, climate change, increase in sea level, and countless other less exposed and less visible problems, are the results of land-use change caused by human activities up to a greater or lesser extent. These environmental problems cause serious impacts like food security, human vulnerability, health and safety, and threaten the overall viability of earth. As Kates et al. (1990) puts it, 'the lands of the earth bear the most visible if not necessarily the most profound imprints of humankind's actions'.

These problems assume even more significant proportions in developing countries like India, which exhibit lower levels of social, economic, and infrastructure amenities compounded with state apathy. Moreover, post-liberalization India has seen more pressing demand for land under conversion on account of export-led and market-driven economic incentives. Among other commodities, marine product exports from India increased substantially after the new economic policies of 1990s, transforming vast areas of land under agriculture and mangrove into aquaculture farmlands. Also, the outbreak of viral disease in shrimp farms in Thailand and Vietnam reduced world supply of the product and provided a big opportunity to India to fill this supply-demand gap.

For effective policy formulations, any analysis of land-use change should answer the question, 'what drives/causes land-use change'. The drivers could be either bio-physical, socio-economic or both. The bio-physical drivers comprise of weather and climate variations, landform, topography, volcanic eruptions, plant succession, soil types and processes, drainage patterns, and availability of natural resources. The socio-economic drivers constitute demographic, social, economic, political and institutional factors and processes such as population and related change, industrial structure and change, technology and technological change, the family, the market, various public sector bodies and the related policies and rules, values, community organization and norms, and property regime.

Keeping this in view, we try to investigate the most important socio-economic factors behind the land-use change pattern in the Indian Sundarbans, West Bengal. The analysis is carried out by combining time-series land-use data, extracted from the high resolution satellite data, and the socio-economic data to estimate the underlying relationship econometrically for the period 1986–2004. The study becomes significant because Sundarbans exhibits a fragile and vulnerable natural system. Though there are various forms of land-use change in the region, but the purpose of the present study is to find out the factors behind the conversion of agricultural and mangrove land to aquaculture land.

EVIDENCE FROM OTHER STUDIES

In several studies across the world, socio-economic data has been combined with the spatial data obtained through remote sensing to understand the drivers of land-use change. Most of these studies primarily attempt to explore the underlying factors of drivers of conversion of land from one use to another.

In a pioneering work inking macroeconomic variables with land-use change, Bilsborrow, and Geores (1995) explore how population change can create pressure on land and hence lead to deforestation (land extensification) and/or more intensification of agriculture by increasing the labour unit per land and reducing fallow times. This in turn leads to land erosion, that is, an environmental degradation. Population pressure may have two sources:

(1) Natural growth: This in turns depend on the following factors:
   (a) Desired Family size and couples' ability to realize such size,
   (b) Education, and
   (c) Urbanization.
(2) Migration: Migration depends on the following factors:
   (a) Availability of untapped lands perceived as potentially productive and unclaimed,
   (b) Whether these lands are accessible by roads,
   (c) Other public policies that stimulate land colonization of new areas, and
   (d) Lack of other alternative attractive destinations.

The study of the accumulated pressure (that is, the net effect of all the factors which either lead to decrease or increase in population) is such that it leads to extensification and/or intensification of agriculture, causes environmental degradation through desertification, and soil erosion. Of course, the effect of population pressure on different areas' environment depends on the ecological characteristics relative to population pressure. An already highly deforested area will have a different marginal effect of a population increase on environmental degradation if further deforestation is carried out than if a similar phenomenon takes place in an area never deforested. But demographic factors are not the only factors that affect intensification and extensification of agriculture; government policy variables are also very important—relative prices of agricultural products, direct and indirect subsidies and taxes, exchange rates for agricultural exports and imports such as agricultural machinery, credit availability and cost, and agricultural research, and dissemination through extension services. Information on all these variables are also required. In fact, if adequate information on these variables are not obtained, then it will be difficult to know the pattern of land-use changes. In different institutional and ecological contexts, these variables have different impact. For example, if due to scarcity of land and the present scenario of environment, the government bans further agricultural extensification or intensification, then it would mean that the present population pressure would have no further effect on environment through agricultural extensification or intensification.

Due to lack of availability of these intermediate data across countries, the authors have adhered to the relationship between demographic, land-use, and environmental variables. The focus is on rural areas of developing countries. Data sources are population census for various years, United Nations estimates, FAO. The main result of the study emerges in the form where there is *absence of any significant correlation between rural population growth and land extensification*. Also, no significant correlation has been observed between changes in land-use deforestation and non-demographic variables. However, given the poor quality of data, these findings are dubious. Land-use data are poor as mentioned by the authors. Definitions of rural and urban areas vary across countries. Data on deforestation are often unreliable, as per the authors and hence suggested satellite imageries. Also cross-country variation in the quality of data has also created unreliable estimates. Infact, the methodology of simple correlation is inadequate because deforestation and land-use changes are explained by many factors, which are unlikely to be represented in simple correlation framework.

Chompitz and Gray (1996) combine remote-sensing data with economic variables and model them with the help of econometric tools. Using satellite data, a multinomial logit model for land-use has been estimated for southern Belize. The land-use categories are natural vegetation (comprising of forest, secondary re-growth wetland, and natural savanna), semi-subsistence agriculture (comprising milpa and other non-mechanized annual cultivation), and commercial agriculture (comprising mostly pasture and mechanized farming). Independent variables are distance to market, land and soil characteristics such as concentration of nitrogen (in per cent), slope (in degrees) of land and phosphorous (available phosphorous in parts per million), a flood hazard dummy, national land (as an indication for encroachment) and zonal forest reserves.

Distance to market was found to be more negatively related to commercial cultivation than to semi-subsistence cultivation, implying that roads are more of a necessity for commercial purpose than for semi-subsistence cultivation. Slope is negatively related to commercial farming but positively with semi-subsistence farming. Also, both types of agriculture have been found to be positively related to soil pH but negatively related to excessively high or low (that is, a negative coefficient of squared pH). The flood hazard dummy is positively related to both types of farming. National land has been found to have low probability of commercial cultivation and high for semi-subsistence cultivation, suggesting that these lands are subject to encroachment. Zonal forest reserve has been found to possess very low relative probabilities of agricultural use. Other independent variables are wetness (defined as an 8-point ordinal scale for drainage ranging from 0 (that is, well drained) to 7 (that is, permanently wet) and rainfall (mean annual rainfall in meters). Wetness and rainfall are negatively related to commercial and positively related to semi-subsistence agriculture. The main conclusion drawn here is that market distance, land quality, and tenure (tenure variables are national land and forest reserves) have strong interactive effect on the likelihood and type of cultivation.

One of the important factors that have been perceived to cause deforestation is increasing population. Pfaff (1999) attempted to understand the causes of deforestation in Amazon—that it is not only the increasing population in a region which leads to deforestation but there are some other variables also which are important, perhaps they can be so important that even if the effect of population is not taken into account, other factors can significantly lead to deforestation. Population, means of communication such as road density (km/sq km country area), river density (km/sq km country area), distance to market, government development project area density (sq km/sq km country area), credit facility (number of credit agencies/ sq km country area), industrial wage and quality of soil (density of nitrogen in soil) are the factors being looked at as the drivers of deforestation, using country-level data for Brazilian Amazon. Land characteristic such as soil quality and vegetation type and factors that affect transport costs such as density of paved roads in a country as well as in the neighbouring countries and distance to major markets are significant. In addition, development project policies appear to have independent effects, although provision of credit infrastructure does not. The important aspect dealt herewith is the finding that population density does not have a significant effect on deforestation when many potential determinants are included.

In another significant methodological study, Cropper, Griffiths, and Mani (1999) focused on the drivers of deforestation and the country of study is Thailand. The dependent variable is the proportion of cleared land. The independent variables are agricultural household density (total agricultural household divided by total area of the province), road density (road divided by total area of the province), slope of land (used as representative of soil quality), acrisol (used as representative of soil quality), distance to Bangkok, price of logs, price and four zonal dummies, viz. Northern Dummy, North-eastern Dummy, Southern Dummy, and Central Dummy. A Linear Probability Model has been estimated with pooled data for the years 1976, 1978, 1982, 1985, and 1989. The estimation suggests that the coefficient of population pressure (agricultural household density) is positive and significant for Thailand as a whole with different degrees of effect on different zones. The coefficient of road density is also positive and significant for Thailand as a whole although a variation in zones is present. The coefficient of distance to Bangkok has been found to be negative and statistically significant, signifying that higher distance from the centre place would lead to lower deforestation because nearer the centre would mean more facilities at lower cost. Log or Timber prices and rice prices are statistically insignificant. The log prices have been included in the model on the basis of the postulate that cost of clearing is reduced by any revenues received from the sale of timber and hence the extent of clearing is assumed to be inversely related with price of logs.

By using spatial econometrics, numerous new insights have been obtained in recent years. The study of Nelson and Hellerstein (1997) falls under this category. This study uses seven categories of land-use as independent variables. These categories are chosen using satellite images of Central Mexico. The authors point out the high possibility of incorrect identification of the land-uses other than forest and irrigated cropland. Therefore, they have not taken into account the other land-uses while estimating the regression equation.

They have used nine explanatory variables. These are six geophysical variables and three socio-economic variables. The geophysical variables are elevation, slope soil, potential intensity of solar radiation, a dummy equal to 1 for flat pixel and a dummy equal to 1 or north-facing slopes. For a region in Central Mexico, they have found that road access to land-use does affect land-use. Roads seem to influence location mostly near currently forested areas. The location identified as forest increases as roads access becomes more difficult. Removing effects of roads from regression allow forests to grow back down mountain sides. Roads also affect positively the irrigated crop areas. Negative effect of roads was found on irrigated cropland. An increase in the transportation cost to roads and villages reduces the probability that a location with irrigated crops will remain cropped and increases the probability that a forested location will remain forested. The effect of transportation cost to the large population centre is also significant on both the categories. In other words, this model also shows that roads are important factors for deforestation.

Another path-breaking study by Nelson, Harris, and Stone (2001) looks at the effect of property right on land-use in three parts of Darien province, Panama for 1997. These three parts are a national park where no human activity is supposed to take place and two reserves for indigenous people. Five categories of land-uses are used as dependent variable. They are forest without Cupio trees, forest with Cupio trees, forest with Cativo trees, Human intervention (livestock pasture, crop land, and brush), and Marsh areas.

These variables influence the productivity of land. To correct the spatial dependence and auto correlation, two spatial lag variables are included. These are lag soil quality and lag slope. Each lag variable is the average of the values of the original variables in the eight pixels surrounding the location. They have included eight socio-economic variables also. Four 0–1 dummies, for Darien national park, concession area (within Darien national park where human activity is permitted) and two reserves, Cemaco and Sambu, are also used as independent variables. Four cost of access variable from land-use locations are used as proxy for price level. The hypothesis is that a plot of land is allocated to that use which has highest net present value. In this context, the theoretical idea behind the use of property right variable (empirically captured through four location dummies) is that property right lowers the discount rate of the operator of a parcel, making long-term investments with larger future payouts more desirable. In other words, this means that providing land-users with more secured property rights will result in more sustainable land-use, preservation of biodiversity, and less deforestation.

However, the study finds that this outcome also depends on the location. The area considered in the study is not accessible by advanced and better mode of transportation and as a result even if after simulating land-uses without taking account of the legal protection to land-use, little difference to land-use pattern is observed. Hence this study points out that it is not solely the property right but location also matters to have effect of human intervention on land-use. The land-use data in this study are satellite imageries.

Kumar (2005) analyses the drivers of land-use change in the context of urban ecosystems of National Capital Region, Delhi. The main focus here is on the macro level socio-economic factors and its effect on the rate at which agricultural land and natural ecosystems are converted to urban uses. An important variable with its obvious global character is influx of foreign capital. This is one of the important variables in the study which has changed the land-use pattern from agricultural use to urban use. It has been strategically hypothesized here that this is one of the important reasons that historically agrarian economy has been transformed to an economy with expanding textile, electronics, and food processing industries. In other words, one important aspect is the study of the effect of globalization through land-use changes.

The study has used two kinds of land-use changes as dependent variable (viz. the land conversion). That is, there are two dependent variables of land conversion;

(1) Natural to urban, and
(2) Agriculture to urban.

Kumar distinguishes between the conversion of agricultural land and non-agricultural land on the basis of the assumption that the opportunity cost of converting agricultural land is greater than the opportunity cost of converting shrub, water or forest. That is, the difference in the opportunity cost of conversion of the two types of land is the basis of demarcation of the two types of land and their conversion.

Hence the finding is the positive effect of foreign direct investment (FDI), especially in construction sector on land conversion to urban uses. The time period covered was 1985 to 2004. The land-use data are satellite imagery. Other data were collected from various official sources.

METHODOLOGICAL FRAMEWORK FOR LAND-USE CHANGE IN THE INDIAN SUNDARBANS

Objective of the Analysis

The main objective of the analysis is to find out the significant (socio-economic) drivers of land-use change in the Indian Sundarbans in the period 1986–2004.

Conceptual Framework

There are two types of models, which discuss the drivers of land-use change:

(i) spatial, and (ii) aspatial.

In the spatial model, location and landscape are in the mainstay. For the aspatial model, the changes in land-use, for example, deforestation, and conversion of productive land to unproductive one, are explained in terms of behaviour and response of land owners. So, the models used here are descriptive in nature, explaining the probable drivers of land-use change rather than predicting any future trends. There could be other types of conversion but their ecological impacts could be either insignificant (ecological impacts of the conversions would be dealt separately) or the magnitude of conversion itself is very small hence ignored in the study. For the conversion, socio-economic determinants are included which cause the conversion from (1) paddy fields to aquaculture farm or, (2) from mangrove to aquaculture farm.

In this chapter, we are trying to understand two types of land-use change in the region of Indian Sundarbans (North and South 24, Paraganas). For the first, the dependant variable is the 'annual cumulative' paddy land in the region that is converted to aquaculture land in block i and year t. In the second case, the variable (Mangrove Forest to Aquaculture farm), would be the annual cumulative mangrove land converted to aquaculture farm in block i in year t,

METHODOLOGY

Panel data analysis endows regression analysis with both a spatial and temporal dimension. The spatial dimension pertains to a set of cross-sectional units of observation. These could be countries, states, counties, firms, commodities, groups of people, or even individuals. The temporal dimension pertains to periodic observations of a set of variables characterizing these cross-sectional units over a particular time span.

The panel data structure confers upon the variables two dimensions. They have a cross-sectional unit of observation, which in this case is block i. They have a temporal reference, t, in this case the year. The error term has two dimensions; one for the country and one for the time period. There are various types of panel models for estimation.

The Constant Coefficients Model

One type of panel model has constant coefficients, referring to both intercepts and slopes. In the event that there is neither significant country nor significant temporal effects, we could pool all of the data and run an ordinary least squares regression model. Although most of the time there are either country or temporal effects, there are occasions when neither of these is statistically significant. This model is sometimes called the 'pooled regression model'.

The Fixed Effects Model (Least Squares Dummy Variable Model)

Another type of panel model would have constant slopes but intercepts that differ according to the cross-sectional (group) unit— for example, the country. Although there are no significant temporal effects, there are significant differences among countries in this type of model. While the intercept is cross-section (group) specific and in this case differs from country to country, it may or may not differ over time. These models are called 'fixed effects' models.

Model Specification

A random effects model, which uses a procedure known as feasible generalized least squares and yields consistent estimates of the coefficients, has been used in the present study, with the following functional form:

$$\log Y_{it} = \alpha + \log \beta X_{it} + \mu_{it} + v_{it}$$

X is the matrix of variables, which are supposed to influence the rate of land-use change. μ is an error term, and

α and β are regression coefficients that are estimated from the panel data with an assumption that β for any block is equal to a mean value ($\bar{\beta}$) for all blocks plus some

$$v_{it} = \alpha_i + \varepsilon_{it}$$

random error ε

where α represents omitted variables that vary across individuals, but not over time; ε denotes omitted variables which vary over time but are constant across individuals.

The log-linear (double log) form of model has been used to make the data approximately normally distributed because of the skewed distribution of some of the variables as they are constructed as ratios. Moreover, the log-linear formulations can take care of the problems of unequal variation and outliers besides making an easy interpretation of the regression results possible.

Conversion of Paddy Land to Aquaculture Farm
Model specification for Conversion 1

log (Paddy Field → Aquaculture Land) =

$\alpha + \beta_1$ log (Value of gross paddy output/paddy workers) + β_2 log (Population density) +

β_3 log [(Value of net paddy output/Paddy land) / (Value of net aqua output/AquaLand)]

In this specification, the conversion of paddy to aquaculture farm is a function related with the labour productivity in paddy field, population density, and the return to paddy land relative to aqualand-uses. Relative land productivity is used as a proxy for the opportunity costs of land conversion. It is quite possible that opportunity cost of conversion of aquaculture is higher than carrying out paddy cultivation. This higher opportunity cost of conversion would induce the greater conversion of paddy fields to aquaculture. Also, if paddy labour productivity is lower than the aqua labour productivity, there will be more incentives to convert paddy land to aquaculture uses. Population density also seems to be a guiding factor behind the conversion under this category.

Conversion of mangrove land to aquaculture farm
Model specification –2 (Log-Linear Formulation)

log (Mangrove → AquaLand) = $\alpha + \beta_1$ log (Population density) +

β_2 log [(Value of net paddy output/paddy land) / (value of net aqua output/aqua land)]+

β_3 log [(Value of gross paddy output/paddy workers) / (value of gross aqua output/aqua workers)]

In this type of conversion, the relative factor productivity (labour) and sectoral productivity ratio would explain the conversion process. In case of both conversions, we would be assuming that the private opportunity cost of converting the paddy field and mangrove areas is far lower than maintaining them as they are (paddy land or mangrove forest). Also, in case of paddy to aquaculture conversion, the differential return of two land-uses is a critical factor.

For remote-sensing data, we would identify the year in which the conversion occurred. This information would be used to generate annual data on land-use change. This can be done with the help of two methods: *Bayesian Maximum Likelihood* (Seto et al. 2002) or *Econometric methods* to identify the year in which land-use changes occurred in a time-series of images. A detailed methodology on remote-sensing based data analysis is given in Appendix 6A.

DATA AND VARIABLES

The analysis of land-use change due to various socio-economic drivers involves the following variables:

Blocks for the analysis:- (N24P): *Sandeshkhali I&II, Minakhan,*
(S24P): *Namkhana, Basanti, Canning I&II, Kakdwip, Gosaba,* and *Kultali.*

For the panel or time-series analysis, ideally, we should have data on most of the variables for all the periods in the study but, unfortunately, we do not have such data on many of the variables for the period 1986–2004 (to make the series continuous between this period, data on land-use has been generated on the basis of NRSA data for the years 1986–9, 1989–96, 1996–2001, and 2001–4, while data on socio-economic variables has been generated on the basis of primary as well as secondary data for different years from different sources.

Table 6.1 shows that blocks Sandeshkhali I&II and Minakhan each experienced conversion of more than 100 sq km of land for aquaculture during the study period while Canning observed about 54 sq km of land being converted to aquaculture.

The study period also witnesses conversion of aquaculture land mainly for settlement purposes. Almost all blocks except Basanti exhibit around 100 per cent land conversion from aquaculture to settlement.

A block-wise trend in land-use transformation from various land-use classifications to aquaculture during 1986 to 2004 is shown in Table 6.2. The biggest chunk of land converted to aquaculture under each block has come from paddy. Since the chapter focuses on land transformation from mangrove (dense forest) and agriculture (paddy land) to aquaculture, therefore, other land transformation categories

Table 6.1: Block-wise Total Land Transformed to and from Aquaculture and Total Land under Aquaculture during 1986–2004.

Blocks	Total Land[1] Transformed to Aquaculture (Sq Km)	Total Land[2] Transformed from Aquaculture (Sq Km)
Sandeshkhali I&II	104.36	17.122 (97.73)
Minakhan	124.41	35.267 (100)
Namkhana	23.72	6.995 (100)
Basanti	37.37	4.538 (77.61)
Canning I&II	54.28	8.502 (100)
Kakdwip	8.16	3.361 (100)
Gosaba	31.91	8.240 (97.31)
Kultali	32.21	5.550 (100)

Source: NRSA (2004).
Note: 1. This column shows total transformed land from various uses for the period.
2. This column shows total transformed land from aquaculture to various uses.
Figures in brackets show percentage of total land converted from aquaculture to settlement.

have not been analysed. Category 'other' comprises of conversion of land from aquaculture to dry aquaculture and from dry aquaculture to aquaculture besides, water bodies, other vegetation, marshy land, mud flats with vegetation, swamp, and reclaimed land from forest. Namkhana and Kultali blocks exhibit conversion of dense forest to aquaculture to the tune of about 25 sq km and 11 sq km each during the study period whereas Minakhan block, with highest land being converted to aquaculture, observes negligible conversions from dense forest. The main chunk of land (about 92 sq km) in Minakhan consisted of categories such as dry aquaculture to aquaculture and vice versa.

Table 6.2: Block-wise Trend in Land-use Transformation from Various Classes to Aquaculture during 1986–2004

Blocks	Transformation from class:	Transformation to Aquaculture (sq km)	Per cent to total transformation
Sandeshkhali	Dense forest	0.384	0.37
	Paddy	56.24	53.82
	Other	41.573	39.83
	Total transformation	104.36	100.00
Minakhan	Dense forest	2.62	2.00.
	Paddy	23.53	19.00
	Other	98.26	79.00
	Total transformation	124.41	100
Namkhana	Dense forest	6.12	25.78
	Paddy	9.82	41.45
	Other	6.70	28.22
	Total transformation	23.72	100.00
Basanti	Dense forest	1.184	3.17
	Paddy	25.79	69.00
	Other	7.53	20.16
	Total transformation	37.37	100.00
Canning	Dense forest	0.67	1.24
	Paddy	25.58	47.13
	Other	25.18	46.39
	Total transformation	54.28	100.00
Kakdwip	Dense forest	0.49	6.04
	Paddy	5.51	67.60
	Other	1.54	18.85
	Total transformation	8.159	100.00
Gosaba	Dense forest	0.71	2.21
	Paddy	21.14	66.23
	Other	7.72	24.19
	Total transformation	31.91	100.00

(Contd.)

(Table 6.2 contd.)

Blocks	Transformation from class:	Transformation to Aquaculture (sq km)	Per cent to total transformation
Kultali	Dense forest	3.50	10.84
	Paddy	15.40	47.82
	Other	11.60	36.02
	Total transformation	32.21	100.00

Source: Compiled by authors based on satellite imageries data from the National Remote Sensing Agency, Hyderabad, India.

Note: Figures for land transformation from paddy to aquaculture have been arrived at by deducting 10 per cent from the total agricultural land transformed to aquaculture, assuming that out of total agricultural land in West Bengal, 90 per cent land is used for paddy cultivation while 10 per cent is used for other crops.

BLOCK-WISE MAPS SHOWING DIFFERENT KINDS OF LAND-USES IN THE STUDY AREA – 2004

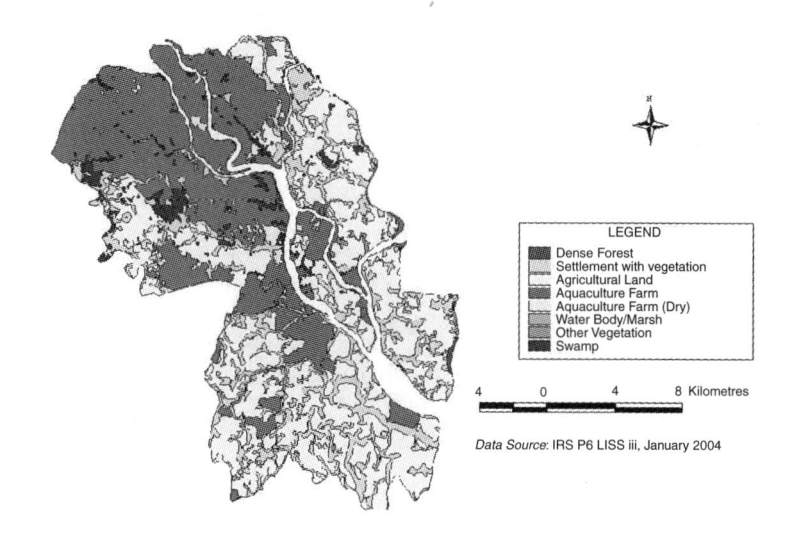

Source: NRSA, 2004.

Map 6.1: Land-use Map of Sandeshkhali (I & II Blocks)

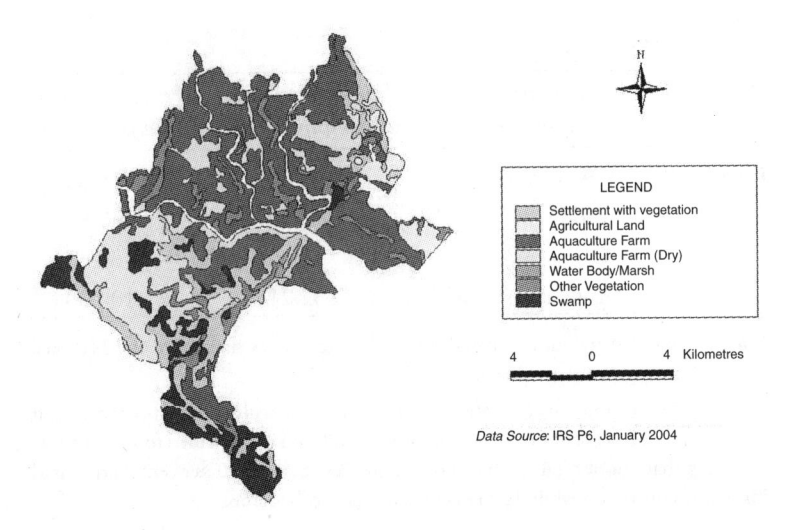

148 Biodiversity, Land-use Change, and Human Well-being

Source: NRSA, 2004.

Map 6.2: Land-use Map of Minakhan Block

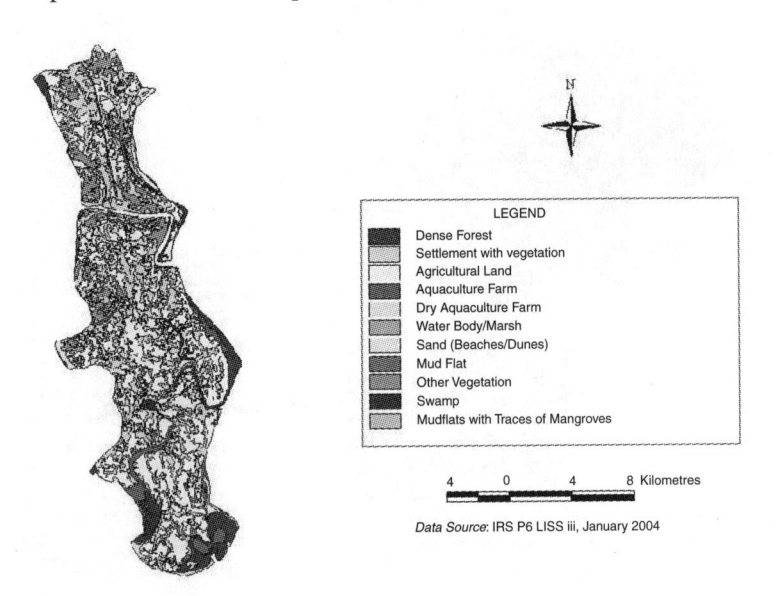

Source: NRSA, 2004.

Map 6.3: Land-use Map of Namkhana

Land-use Change in the Sundarbans 149

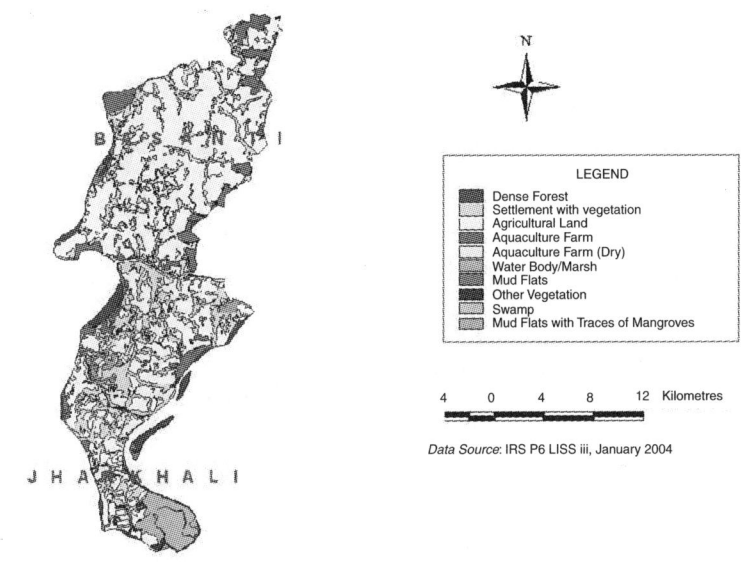

Source: NRSA, 2004.

Map 6.4:   Land-use Map of Basanti

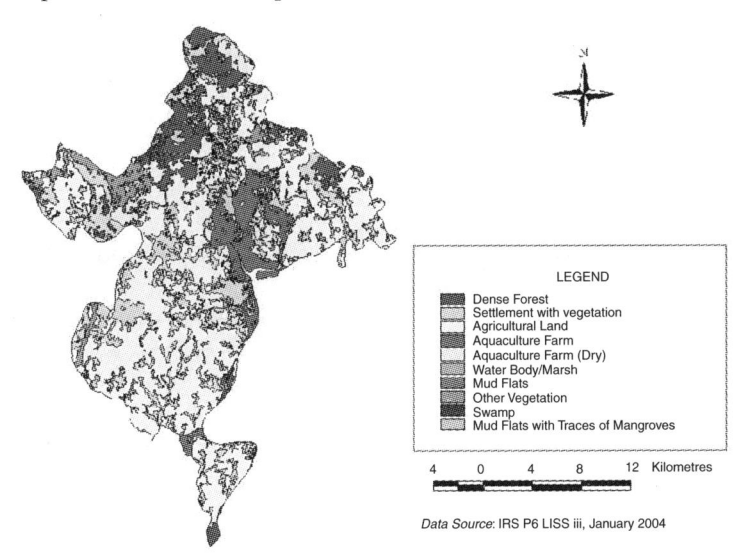

Source: NRSA, 2004.

Map 6.5:   Land-use Map of Canning I & II

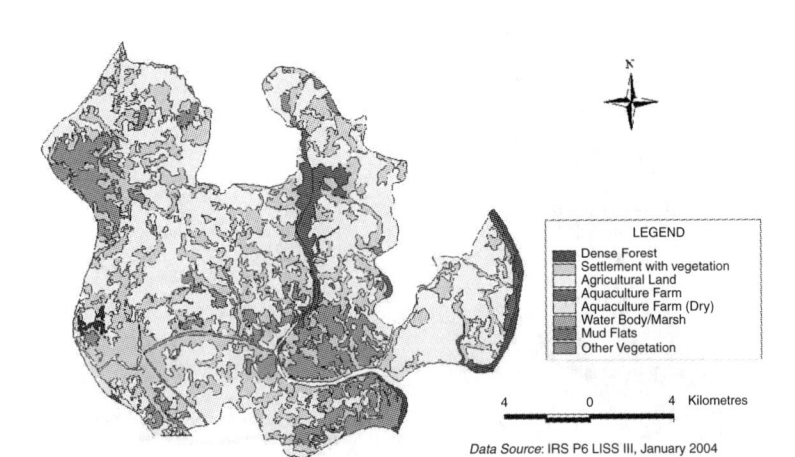

150 Biodiversity, Land-use Change, and Human Well-being

Source: NRSA, 2004.

Map 6.6:   Land-use Map of Kakdwip

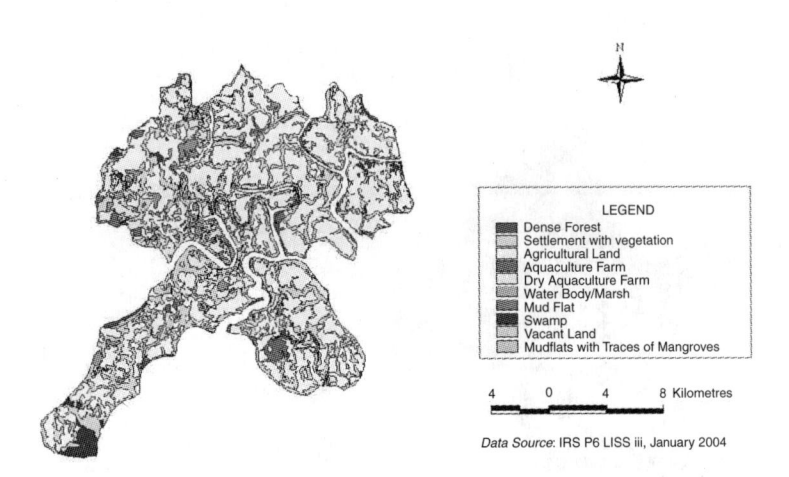

Source: NRSA, 2004.

Map 6.7:   Land-use Map of Gosaba

Land-use Change in the Sundarbans 151

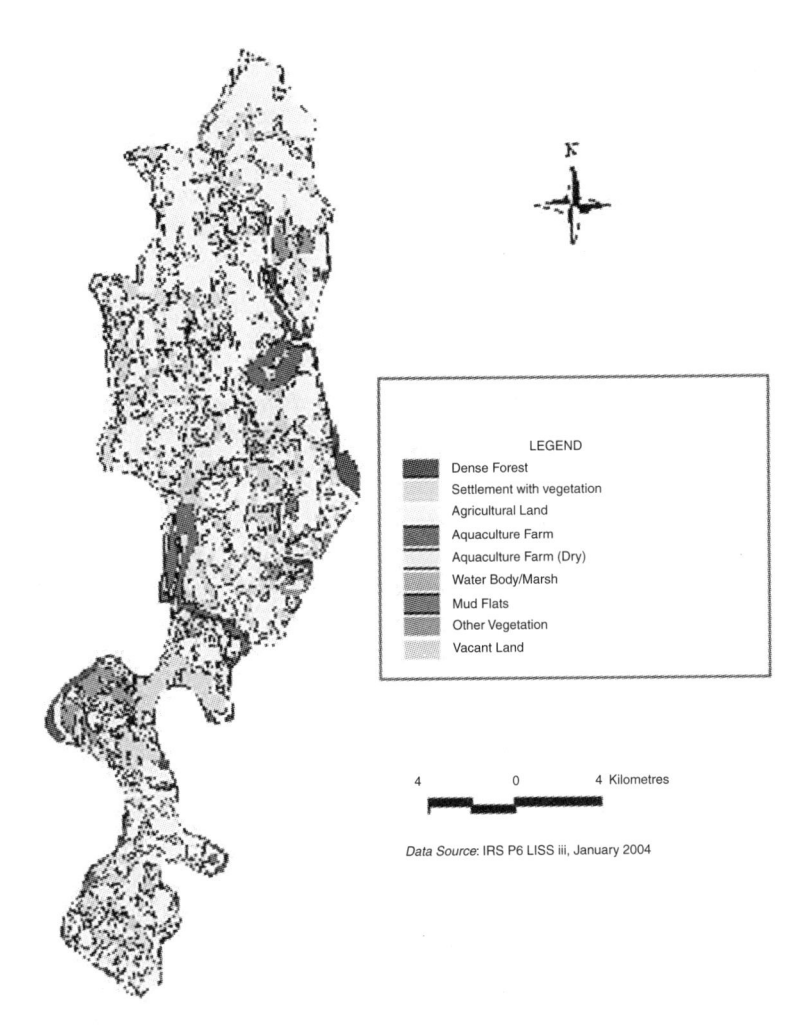

LEGEND
Dense Forest
Settlement with vegetation
Agricultural Land
Aquaculture Farm
Aquaculture Farm (Dry)
Water Body/Marsh
Mud Flats
Other Vegetation
Vacant Land

Data Source: IRS P6 LISS iii, January 2004

Source: NRSA, 2004.

Map 6.8:   Land-use Map of Kultali

DESCRIPTION AND CONSTRUCTION OF VARIABLES

Dependent Variables

PDGLD—Annual absolute (cumulative) paddy land converted to aquaculture

MNGL—Annual absolute (cumulative) mangrove (dense forest only) land converted to aquaculture

(Note: Data on dependent variables is in absolute cumulative figures making it compatible with the independent variables since the latter do not show percentage change over the years but absolute cumulative changes. For example, change in population density from one year to next is a cumulative figure having both the persons in the previous period as well as the new additions in the next period).

Independent Variables

POPDEN—Population density

NETRELLDPROD—(Value of net paddy output/paddy land)/(value of net aqua output/aqua land)

RELLBPROD—(Value of gross paddy output/paddy workers/(value of gross aqua output/aqua workers)

PDLBPROD—Value of gross paddy output/paddy workers

MNGLD—(Annual absolute (cumulative) mangrove (dense forest only) land converted to aquaculture) (Ha): This variable has been constructed by using NRSA data on land conversion from dense forest to aquaculture in the periods 1986–9, 1989–96, 1996–2001, and 2001–4 for all the blocks of the study. To make data continuous for the period 1986–2004, period-wise averages were calculated by dividing the data under each period by number of years in that period to arrive at the annual data. From the first period, that is, 1986–9 onwards, for every next period, the previous land conversion has been added to the next period to make the data continuous and compatible with the independent variables.

PDLD—(Annual absolute (cumulative) paddy land converted to aquaculture) (Ha): This variable has been constructed by using NRSA data on land transformation from paddy land (since data on paddy land is not given in the NRSA data, therefore, to arrive at paddy land converted to aquaculture, 90 per cent of land converted from agriculture to aquaculture has been used, assuming that 10 per cent of this land is used for other vegetation) to aquaculture in the periods 1986–9, 1989–96, 1996–2001, and 2001–4 for all the blocks of the study. To make data continuous for the period 1986–2004, period-wise averages were calculated by dividing the data under each period by number of years in that period to arrive at the annual data. Here also, from the first period, that is, 1986–9 onwards, for every next period, the previous land conversion has been added to the next period to make the data continuous and compatible with the independent variables.

AQLD—(Aquaculture land) (Ha): This variable constitutes total aquaculture farm land (including dry aquaculture farms) for each block. Data has been generated for the period 1986-2004 on the basis of NRSA data for the years 1986, 1989, 1996, 2001, and 2004.

Avg. Aqua—(Avg. value of aquaculture) (Rs/ha): This variable has been constructed from the primary survey data of aquaculture farms for the period 2000–4. Average per hectare value for Minakhan, Canning, and Gosaba blocks were calculated from the survey data and then figures for the series 1986–2004 have been calculated from the linear interpolation of the available data. This generated data was filled for the remaining five blocks of the study according to the districts under which different blocks fall.

VALAQ—(Value of aquaculture production (Rs): This variable has been arrived at by multiplying average aquaculture production (Avg. Aqua) by aquaculture land (AQLD) for each block and for each year of the study.

POPDEN—(Population density) (Persons/sq km): This variable is based on the data from District Census Handbooks and Village and Town Directories, Census of India, 1991 and 2001. First of all, on the basis of 1991 and 2001 figures, population for each block was generated for all the years of the study with the help of compound growth rate method and then population (all persons) for every block in the study was divided by area (sq km) under each block to arrive at the density at block level.

PADPOP—(Paddy cultivation workers): This variable is based on the data from District Census Handbooks, Census of India, 1991 and 2001. First, all the persons involved in agricultural activities (main and marginal cultivators and labourers) were taken together for each block to get the total number of persons involved in agriculture. Secondly, 10 per cent of this population was subtracted to arrive at total persons engaged in paddy cultivation. Finally, data was generated by linear interpolation for the study period 1986–2004 for each block separately.

AQPOP—(Aquaculture workers): This variable has been arrived at from the PADPOP variable. Out of the total paddy workers, 10 per cent was assigned to aquaculture workers.

Avg. Paddy—(Avg. value of paddy production) *(Rs/ha):* This variable has been constructed from the secondary data for N24P and S24P from many sources, such as, Evaluation Wing, Directorate of Agriculture, Govt. of West Bengal, *Economic Review* 2001; *Reports of the Commission for Agricultural Costs and Prices*, Dept. of Agriculture and Cooperation, Ministry of Agriculture, Govt. of India, 2001; Bureau of Applied Economics, and Statistics 2002 and Past Issues. Data is available for the period 1986–2001 for districts but we do not have this data at the block level, so, data for 2001–4 has been extrapolated using compound growth formula. In order to fill the data under blocks, we have used the data for N24P to blocks falling under this block whereas data for S24P has been used for the blocks under S24P.

PDLD1—(Paddy land) (Ha): Using the NRSA data, this variable has been constructed by deducting 10 per cent from the total agricultural land under each block to constitute total paddy land (assuming that out of total area under agriculture in these blocks 90 per cent is used for paddy cultivation). Data has been generated for the period 1986–2004 on the basis of NRSA data for the years 1986, 1989, 1996, 2001, and 2004.

VALPD—(Value of paddy production) (Rs): This variable has been arrived at by multiplying average paddy production (Avg. Paddy) by paddy land (PDLD1) for each block and for each year of the study.

EMPIRICAL ESTIMATION OF THE LAND-USE CHANGE

Conversion of Paddy Land to Aquaculture (Log-linear Specification)

LogCUMPDLD = f (LogPOPDEN, LogNETRELDPROD, LogPDLBPROD)- (1)

CUMPDLD—Annual absolute (cumulative) paddy land converted to aquaculture

POPDEN—Population density

NETRELLDPROD—(Value of net paddy output/paddy land)/(value of net aqua output/aqua land)

PDLBPROD—Value of gross paddy output/paddy labourers

Breusch-Pagan Lagrangian Multiplier (LM) test of the significance of random effects confirms that the total variance is better treated as

Table 6.3: Conversion of Paddy Land to Aquaculture (Random Effects Model)

Variable	Coefficients
Log POPDEN = Population density	0.401 (1.65)
Log NETRELDPROD = (Value of net paddy output/paddy land)/(value of net aqua output/aqua land)	–0.479* (–7.44)
Constant	1.357 (0.92)
Breusch-Pagan Lagrangian Multiplier (LM) Chi^2 (Probability LM Chi^2)	794.45 (0.000)
Hausman Chi^2 (Probability Hausman Chi^2)	2.46 (0.293)
Adjusted R^2	0.16
N (Number of groups = 8, Observations per group = 19)	152

Source: Authors' calculations.
Notes: * Significant at 5 per cent level of significance
z statistics are in the parenthesis

components of variation; both within and between individuals (Chi^2= 794.45, p = 0.000). This result tells us that the fixed effects model with a single constant term is inappropriate for our data. The test rejects the null hypothesis in favour of the random effects model.

For testing the specification of the model for our study, another test, the Hausman test statistic Chi^2 with k degrees of freedom (here k=2) gives a value of 2.46. The critical value from the Chi^2 table with two degrees of freedom is 5.99, which is larger than the test statistic. The hypothesis that the individual effects are uncorrelated with the other regressors in the model cannot be rejected. The test proposes that these effects are uncorrelated with the other variables in the model. LM test proves that there are individual effects and the Hausman test suggests that these effects are uncorrelated with the other variables in the model; hence we prefer the random effects specification.

Initially, we started with three explanatory variables in the study but the variable paddy labour productivity had high negative correlation with another explanatory variable population density. Therefore, it was dropped to avoid the multi-collinearity problem.

As shown in Table 6.3, the results of the paddy to aquaculture conversion show statistically significant relationships of net relative land productivity and population density with the paddy land conversion to aquaculture. The sign of the coefficient of net relative land productivity (of paddy relative to aquaculture) is negative, which is a reliable estimate and it tells us that if the yield from paddy goes down, there will be more incentive for that land to be converted to aquaculture, otherwise not. Moreover, the magnitude of –0.47 on the variable tells us that, on an average, there will be conversion of 0.47 per cent of land from paddy to aquaculture as a result of 1 per cent decrease in the relative yield ratio of paddy to aquaculture. The coefficient of the population density shows positive sign implying that as the overall population pressure (including natural growth net of migration) on the blocks increases, land under paddy goes for the conversion to aquaculture. This can be explained through the higher labour productivity and greater employment opportunities available in the aquaculture activity. The coefficient shows a size of 0.40, which means that for every 1 per cent increase in population per sq km, there will be a corresponding conversion of 0.40 per cent of land from paddy to aquaculture.

CONVERSION OF MANGROVE LAND TO AQUACULTURE
(LOG-LINEAR SPECIFICATION)

LogCUMMNGLD = f (LogPOPDEN, LogNETRELLDPROD, RELLBPROD)- (2)

CUMMNGLD—Annual absolute (cumulative) mangrove (dense forest only) land converted to aquaculture

POPDEN—Population density

NETRELLDPROD—Value of net paddy output/paddy land)/(value of net aqua output/aqua land)

RELLBPROD—(Value of gross paddy output)/(paddy workers) / value of gross aqua output/aqua workers)

Table 6.4: Conversion of Mangrove Land to Aquaculture

Variable	Coefficients
log POPDEN = Population density	0.551*
	(2.57)
log NETRELDPROD = (Value of net paddy output/ paddy land)/(value of net aqua output/aqua land)	−0.119*
	(−2.34)
log RELLBPROD - (Value of gross paddy output)/(paddy workers) / value of gross aqua output/aqua workers)	−0.407*
	(−4.41)
Constant	−1.92
	(−1.32)
Breusch-Pagan Lagrangian Multiplier (LM) Chi^2	835.16
(Probability LM Chi^2)	(0.000)
Hausman Chi^2	5.75
(Probability Hausman Chi^2)	(0.1242)
Adjusted R^2	0.059
N (Number of groups = 7, Observations per group = 19)	133

Source: Authors' calculations.
Notes: * Significant at 5 per cent level of significance.
    z statistics are in the parenthesis.

Table 6.4 reports the results of mangrove to aquaculture conversion. In this type of conversion too, Breusch Pagan Lagrangian Multiplier (LM) as well as the Hausman specification tests validate the use of a random effects model rather than the fixed effects model. The LM test statistic of 835.16 is larger than the critical value of Chi^2 at 1 per cent significance level, with one degree of freedom (k-1). The Hausman test statistic Chi^2 with 2 degrees of freedom gives a value of 5.75. The critical value from the Chi^2 table with two degrees of freedom is 5.99, which is higher than the test statistic. Based on the LM and Hausman tests, we use random effects as a superior option.

In case of mangrove to aquaculture conversion, population density, net relative land productivity, and relative labour productivity (productivity of paddy labour relative to aquaculture labour) variables illustrate statistically significant relationship with the land conversion. Here too, the sign of the coefficient of net relative land productivity is negative, which is a consistent estimate and it tells us that there will be conversion of mangrove land to aquaculture land, given that the yield from the paddy relative to aquaculture decreases. This variable also highlights an interesting finding that the land under mangrove can be directly converted to aquaculture uses. The magnitude of this variable is –0.11, which says that, if the yield from paddy relative to aquaculture drops by 1 per cent, there will be conversion of 0.11 per cent of land from mangrove to aquaculture.

The variable relative labour productivity presents a negative relationship with land-use change entailing that if paddy labour productivity is lower relative to aquaculture labour productivity, there is greater incentive to convert mangrove land to aquaculture activities. In case of higher relative paddy productivity, paddy land generates higher proceeds than aquaculture land leading to less motivation to convert mangrove land to aquaculture land. Otherwise, mangrove areas may be more pressed for conversion to reap higher profits on account of relative high labour productivity in paddy. The size of this variable is –0.40 meaning that for every 1 per cent fall in relative labour productivity ratio, there will be 0.40 per cent of land conversion from mangrove to aquaculture.

Population density comes out to be the major variable in this kind of land-use change. The variable shows a positive relationship with mangrove land conversion. On an average, 0.55 per cent of mangrove land goes to aquaculture in response to 1 per cent increase in population density in the blocks.

CONCLUDING REMARKS

Land-use change in the Sundarbans, especially loss of mangrove forest, has been one of the key concerns for various stakeholders—as these forests are critical to the existence of rich biodiversity in the region besides its various ecological functions beneficial to the people. Shrinking forest takes through the route of conversion of mangrove forest for other uses like agriculture and aquaculture. Before we began our exploration of this aspect of mangrove loss, there has been a conflicting view coming from different advocacy groups (saying there has been tremendous loss of forest) and the forest officials (saying there is no loss to mangrove forest). Our analysis of land-use satellite data specially procured from the National Remote Sensing Agency, Department of Space, Government of India for 19 years (1986–2004) for the same season clearly shows the decline in the dense mangrove forest. However, the loss through conversion is not as high as usually claimed. Other conversions for example, from agriculture to aquaculture is far more pronounced than any type of conversion recorded in the data analysis. A behavioural model of conversion of mangrove to aquaculture and agriculture to aquaculture highlights the reasons in terms of encroachment of people in the fragile areas (population density), differentials in return on different land-use and differential in productivity of people on different types of activities. While considering the returns, mangrove forest has been reported to be having no return or insignificant return in the official statistics. This is arising due to the perennial problem of unaccounting of ecological services provided by mangrove forests. This 'externality' effect induces people to convert the land for other purposes that show relatively higher return. Illegal encroachment of the forest seems to be another important factor (not captured in the analytical model) for loss of mangrove. The Government here can either design fiscal mechanism to control the conversion (charges etc.) or provide subsidy to the farmers enabling the existing land more lucrative and persuading them not to convert their forestland for aquaculture or agriculture. Strict enforcement of forest legislation also remains a useful tool.

APPENDIX 6A

METHODOLOGY FOR REMOTE SENSING BASED DATA ANALYSIS

6A.1: Methodology

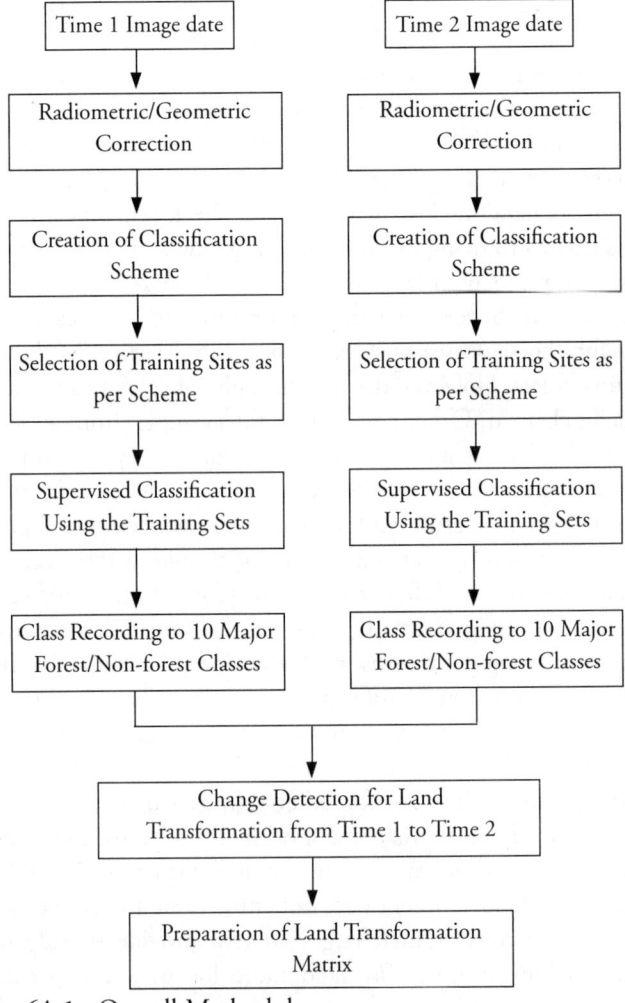

```
Time 1 Image date                    Time 2 Image date
       ↓                                    ↓
Radiometric/Geometric                Radiometric/Geometric
     Correction                           Correction
       ↓                                    ↓
Creation of Classification           Creation of Classification
       Scheme                              Scheme
       ↓                                    ↓
Selection of Training Sites as       Selection of Training Sites as
       per Scheme                          per Scheme
       ↓                                    ↓
Supervised Classification            Supervised Classification
Using the Training Sets              Using the Training Sets
       ↓                                    ↓
Class Recording to 10 Major          Class Recording to 10 Major
Forest/Non-forest Classes            Forest/Non-forest Classes
                       ↓
        Change Detection for Land
        Transformation from Time 1 to Time 2
                       ↓
        Preparation of Land Transformation
                    Matrix
```

Figure 6A.1: Overall Methodology

Source: Hazra and Mitra 2005.

DELIVERABLES

- Study and analysis of land-use/land cover characteristics of the habited parts and forest covered areas of the Indian Sundarbans for 1986, 1989, 1995, 2001, and 2004.
- Change Detection Analysis of different land-use patterns of the habitated parts and forest-covered areas of the Indian Sundarbans from the years 1986–9; 1989–96; 1996–2001, and 2001–4.
- Documentation of the land-use/land cover study for habited parts and forest covered areas of the Indian Sundarbans for the years 1986, 1989, 1995, 2001, and 2004 in word format.
- Documentation of the Change Detection Analysis of different land-use patterns of habitated parts and forest-covered areas of the Indian Sundarbans for the years 1986–9; 1989–96; 1996–2001, and 2001–4 in word format.
- Land-use/land cover maps for the habitated parts and forest covered areas of the Indian Sundarbans for the year 1986, 1989, 1995, 2001, and 2004.

DATA PRODUCTS USED

Different sets of data for generation of land-use/land cover maps of Indian Sundarbans consisting of 13 blocks in South 24 Parganas and 6 blocks in North 24 Parganas. The data are the following:

- LANDSAT TM, January 1986
- IRS 1A LISS I, January 1989
- LANDSAT TM, February 1996
- IRS 1D LISS III, January 2001
- IRS P6 LISS III, January 2004

SOFTWARE USED

- ERDAS IMAGINE (Version 8.5)
- ARCVIEW (Version 3.2A)

IMAGE CLASSIFICATION

The overall objective of the image classification procedure is to sort all the pixels of an image into a finite number of individual land cover classes or themes. Fundamentally, spectral classification forms the bases to objectively map the areas of the image that have similar spectral reflectance/emissivity characteristics. Depending on the type of information required, spectral classes may be associated with identified features in the image (supervised classification) or may be chosen statistically (unsupervised classification).

Normally, multispectral data is used to perform classification and the spectral pattern present within the data for each pixel is used as the numerical basis for categorization. Different feature types manifest different combinations of DNs based on their inherent spectral reflectance and emittance properties. *Pattern* refers to the set of radiance measurements obtained in the various wavelength bands for each pixel. *Spectral pattern recognition* is the family of classification procedures that utilizes this pixel-by-pixel spectral information as the basis for automated land cover classification.

Spatial pattern recognition involves the categorization of image pixels on the basis for their spatial relationship with pixels surrounding them. Spatial classifiers might consider such aspects as image texture, pixel proximity, feature size, shape, directionality, repetition, and context. These types of classifiers attempt to replicate the kind of spatial synthesis done by the human analyst during the visual interpretation process. Accordingly, they tend to be much more complex and computationally intensive than spectral pattern recognition procedures.

Temporal pattern recognition uses time as an aid in feature identification. In agricultural crop surveys, for example, distinct spectral and spatial changes during a growing season can permit discrimination on multi-date imagery that would be impossible given any single date. An interpretation of imagery from either date alone would be unsuccessful, regardless of the number of spectral bands. The data was for the month of January and February for analysing the land-use pattern of the Sundarbans because that is the harvesting season in West Bengal and the crops during this time are cut off and the fields are empty so that the discrimination between the settlement with vegetation and crops are easily done.

Supervised Classification is closely controlled by the analyst. In this process, pixels representing patterns or land cover features are selected, or identified with help from other sources such as aerial photos, ground truth data, or maps. Knowledge of the data, and of the classes desired, is required before classification. By identification of patterns, the computer system is instructed to identify pixels with similar characteristics.

Unsupervised Classification is more computer-automated. It enables the specification of some parameters used by the computer to uncover statistical patterns that are inherent in the data. These patterns do not necessarily correspond to directly meaningful characteristics of the scene, such as contiguous, easily recognized areas of a particular soil type of land-use. They are simply clusters of pixels with similar spectral characteristics.

Unsupervised training is depended upon the data itself for the definition of classes. This method is usually used when less is known about the data before classification. It is then the analyst's responsibility, after classification, to attach meaning to the resulting classes.

Both supervised and unsupervised procedures are applied in two separate steps. The difference between both is that supervised classification involves a training step followed by a classification step, whereas in the unsupervised approach image, data are first classified by aggregation into the natural spectral groupings (clusters) present in the scene, then image analyst determines land cover identity of these spectral groups by comparing the classified image data to ground reference data.

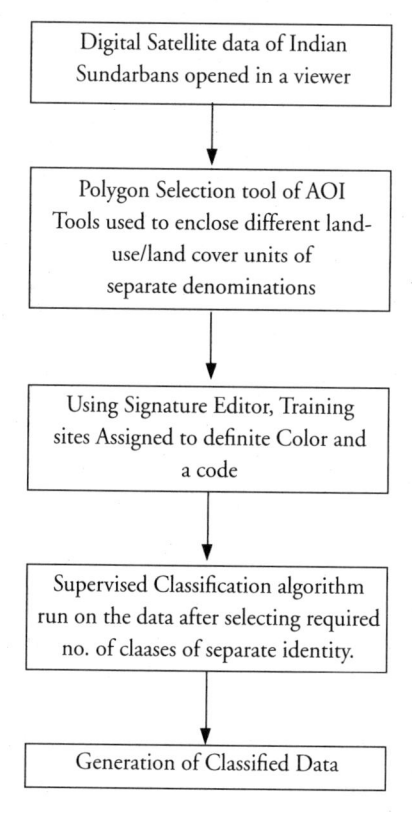

```
┌─────────────────────────────────┐
│   Digital Satellite data of Indian  │
│   Sundarbans opened in a viewer │
└─────────────────────────────────┘
              │
              ▼
┌─────────────────────────────────┐
│   Polygon Selection tool of AOI │
│   Tools used to enclose different land- │
│   use/land cover units of       │
│   separate denominations        │
└─────────────────────────────────┘
              │
              ▼
┌─────────────────────────────────┐
│   Using Signature Editor, Training │
│   sites Assigned to definite Color and │
│   a code                        │
└─────────────────────────────────┘
              │
              ▼
┌─────────────────────────────────┐
│   Supervised Classification algorithm │
│   run on the data after selecting required │
│   no. of claases of separate identity. │
└─────────────────────────────────┘
              │
              ▼
┌─────────────────────────────────┐
│   Generation of Classified Data │
└─────────────────────────────────┘
```

Figure 6A.2: Database creation of Supervised Classification (Raster based)

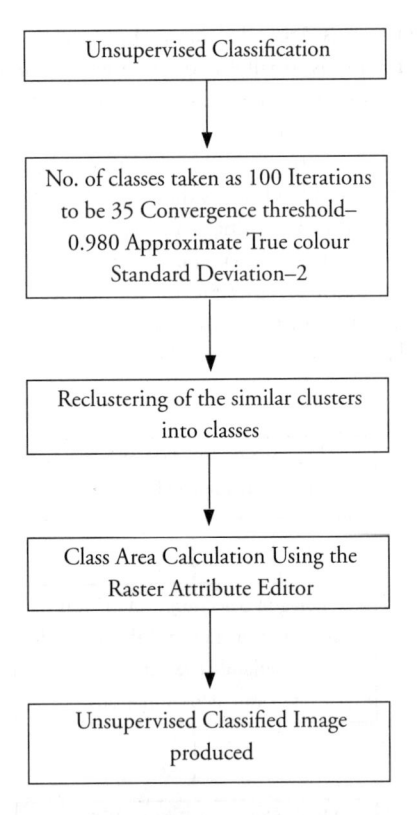

```
┌─────────────────────────────┐
│  Unsupervised Classification │
└──────────────┬──────────────┘
               ▼
┌─────────────────────────────┐
│ No. of classes taken as 100 │
│ Iterations to be 35         │
│ Convergence threshold–      │
│ 0.980 Approximate True      │
│ colour Standard Deviation–2 │
└──────────────┬──────────────┘
               ▼
┌─────────────────────────────┐
│ Reclustering of the similar │
│ clusters into classes       │
└──────────────┬──────────────┘
               ▼
┌─────────────────────────────┐
│ Class Area Calculation Using│
│ the Raster Attribute Editor │
└──────────────┬──────────────┘
               ▼
┌─────────────────────────────┐
│ Unsupervised Classified     │
│ Image produced              │
└─────────────────────────────┘
```

Figure 6A.3: Database creation of Unsupervised Classification (Raster-based) Classification

6A.6   DATABASE CREATION

Importing the Data: The digital satellite data procured from NRSA, Hyderabad, was first imported in Erdas Imagine 8.5 from Generic Binary to software compatible format. The raw data supplied in CD was copied to the hard disk of the computer in a definite file and using the IMPORT option of the ERDAS Imagine, data in GENERIC BINARY format was selected from the product folder and transferred to the destination file format of imagine.

In import data options dialogue–data description, data format was specified as BIL format, and data type as unsigned 8 bits. In the present case of Landsat TM data as per the header file information, the image-recorded length was given to 6700, the number of rows and columns were specified as 5681 and 6120 respectively. The number bands were taken as 7 and file header bytes as 540. In this way, all the sets of data were imported from the generic binary format.

Data Pre-processing: The data was referenced with the option of Geometric Correction from the Raster of the viewer in which the data is opened, the polynomial model properties dialog box opened where polynomial order is taken as I where the output map projection was defined. The projection was set from GCOP too. GCP tool reference was set up by collection reference points from the existing viewer, viewer selection instructions dialog box was opened, and the reference coordinates were taken from reference map information.

The referenced data was coordinated with ground coordinate system and the data got ready for digitization. The forest-covered part of the Indian Sundarbans was cut out from the full referenced data by the process called 'subsetting'. At first the data was opened in any empty viewer and by selecting the polygon tool from the AOI tools of the viewer, the area of interest was marked and thus all the different islands of Indian Sundarbans were separated into different files, such as Lothian, Dhanchi, Susnichara, Ajmalmari East, Ajmalmari West, Ajmalmari North West, Dulibhasani East, Dulibhasani West, Herobhanga, Bulchery, Bhangaduni, Dalhousie, Gosaba, Matla, Saznekhali North, Saznekhali South, Jhilla, Katuajpuri, and Jambudwip.

The human habitated parts of the Indian Sundarbans which were digitized in separate layers are: Sandeshkhali, Haroa, Hasnabad, Hingalaganj north, Hingalganj south, Minakhan, Sagar, Ghoramara, Kakdwip, Patharpratimamainland, Basanti mainland, Jharkhali, Bhagabatpur, Canning, Gurguria, Jayanagar, Chotamullakhali, and Jhinga Abad. Other habitated parts include Kamakhyapur, Kultali mainland, Kumiramari, Maheshpur, Mathurapur, Mitrabari, Moushuni, Namkhanamainland, Namkhanaisland, Paschim Sripatinagar, Pinhali Abad, Rakhalpur, Ramnagar, Satyanarayanpur, Shibnagar, and Dakshin Surendranagar. Later these layers have been arranged according to blocks.

The land-use classes, adopted for the Indian Sundarbans for the human habitated parts are the following:

1. Dense Forest
2. Settlement with Vegetation
3. Agriculture land
4. Aquaculture farm

5. Aquaculture farm (Dry)
6. Water body/ Marsh
7. Sand (Beaches/ Dunes)
8. Mud Flats
9. Other vegetation
10. Swamp
11. Vacant Land
12. Reclaimed land from forest, and
13. Mudflats with traces of mangroves

The land-use classes, adopted for the Indian Sundarbans for the forest-covered parts are the following:

1. Dense Forest
2. Degraded Forest
3. Saline Blanks
4. Water Body
5. Sand (Beaches/ Dunes)
6. Mud flats, and
7. Reclaimed land

The forest-covered area of the Sundarbans has been classified in ERDAS by adopting the Unsupervised Classification. After classification, the area have been calculated in that software itself. Some field observations were taken into account while validating the land-use units. So some field photographs are attached along with this chapter.

ERDAS IMAGINE uses the ISODATA algorithm to perform an unsupervised classification.

The ISODATA (Interactive Self-Operational Data Analysis Technique) algorithm was adopted for doing the classification for generating the different land cover classes on the basis of DN values before field verification. For initializing from Statistics, the checkbox was generated for arbitrary clusters from the file statistics. Number of classes was taken at a range between 60 to 100 and they were grouped.

The ISODATA utility repeats the clustering of the image until either: A maximum number of iterations had been performed, or a maximum number of unchanged pixels had reached between two iterations:

Maximum Iterations: They refer to the maximum number of times the ISODATA utility should re-cluster the data. These parameters prevent the utility form running too long, or from potentially getting stuck in a cycle without reaching the convergence threshold.

Convergence Threshold: This specifies the convergence threshold, which is the maximum percentage of pixels whose cluster assignments can go unchanged between iterations. This threshold prevents the ISODATA utility from running indefinitely the default values for the above. By specifying a convergence threshold of 0.98, one would specify that as soon as 95 per cent or more or the pixels stay in the same cluster between one iteration and the next, the utility should stop processing. In other words, as soon as 5 per cent or fewer of the pixels change clusters between iterations, the utility will stop processing. Then the programme was executed. By opening the unsupervised true-coloured the Raster Attribute Editor was opened to change the colour scheme.

6A.7 CHANGE DETECTION PROCEDURE IN ERDAS ENVIRONMENT

After the classification of the image 1 and image 2 data, the classification scheme, which consisted of 100 classes at first, was recoded into 9 decided classes. By recoding of an image means the recode option has been set up in the Spatial Model Maker in the Raster platform.

6A.7.1 Function Definition

Objects used in a Model Maker model are operated upon with function definitions that one writes with the Model Maker. The function definition is an expression (such as 'a + b + c') that defines your output. You will use a variety of mathematical, statistical, Boolean, neighbourhood, and other functions plus the input object that you set up to write a function definition.

Using the Function Definition Dialog, the Performed Matrix Overlay for the image 1 and image 2 data, a thematic layer is produced, which contains a separate class for every coincidence of classes in two layers. Coding of individual classes for each coincidence is done by:

Table 6A.1: Coding of Individual Classes

Image 1 Time 1		Image 2 Time 2	
Class	Code	Class	Code
Agriculture land	3	Aquaculture farm	4

Source: Hazra and Mitra 2005.

Table 6A.2: True Coding of Individual Classes

Image 1 Time 1		Image 2 Time 2		True Code
Class	Code	Class	Code	Transformed Code
Agriculture land	3	Aquaculture farm	4	3*10+4

Source: Hazra and Mitra 2005.

Classification of transformation code is then given as 34 for the conversion of agriculture land to aquaculture farm while the no change is given as the code 100, erosion 200, and accretion 300. There is recoding of individual classes into defined transformation classes.

Table 6A.3: Preparation of Changed Matrix

Transformation	Area
Agriculture land to aquaculture farm	36.587 $(kms)^2$
Dense Forest	6.324 $(kms)^2$

Source: Hazra and Mitra 2005.

The human habitated part of the Sundarbans was digitized in the Arc View environment by selecting the line format. This line format was used for enclosed polygon, which was then cleaned and built in the ERDAS environment for restoration of the topology. The topology is the spatial relationship between connecting or adjacent feature (for example, arcs, nodes, polygons, and points.) For example, the topology of an arc includes its form- and –to nodes and its left and right polygons. Topological relationships are built from simple elements into complex elements: points (simplest elements) and arcs (set of connected arcs). Redundant data (coordinates are eliminated because an arc) may either represent a liner feature, part of the boundary of an area feature or both. N Topology is useful in GIS because many spatial modelling operations don't require coordinates, but only topological information. The data then was ready for database preparation. A column was added in the database so that ids of respective land-use classes could be imparted to get the land-use/ land cover maps. The *.arc info coverage format was converted to shape-files so that any kind of further changes can be incorporated in the database. In this way, the database was ready and exported to dbase format so that it can open in any software, such as Microsoft Excel, Microsoft Word etc. The methodology for the change detection analysis, which was carried out for the human habited part of the Indian Sundarbans, is as follows:

```
┌─────────────────────────┐         ┌─────────────────────────┐
│ Land-use/Land cover     │         │ Land-use/Land cover     │
│ layer of Patharpratima— │         │ layer of Patharpratima— │
│ Year 1986               │         │ Year 1989               │
└───────────┬─────────────┘         └──────────┬──────────────┘
            │                                  │
            └──────────►  ╭─────────────╮ ◄────┘
                          │ Geoprocessing│
                          │ Wizard Union │
                          │ Option       │
                          ╰──────┬──────╯
                                 ▼
                    ┌────────────────────────┐
                    │ Change Detection       │
                    │ Analysis to get the    │
                    │ land transformation    │
                    │ matrix                 │
                    └───────────┬────────────┘
                                ▼
                    ┌────────────────────────┐
                    │ Generation of Land     │
                    │ Transformation Map     │
                    └────────────────────────┘
```

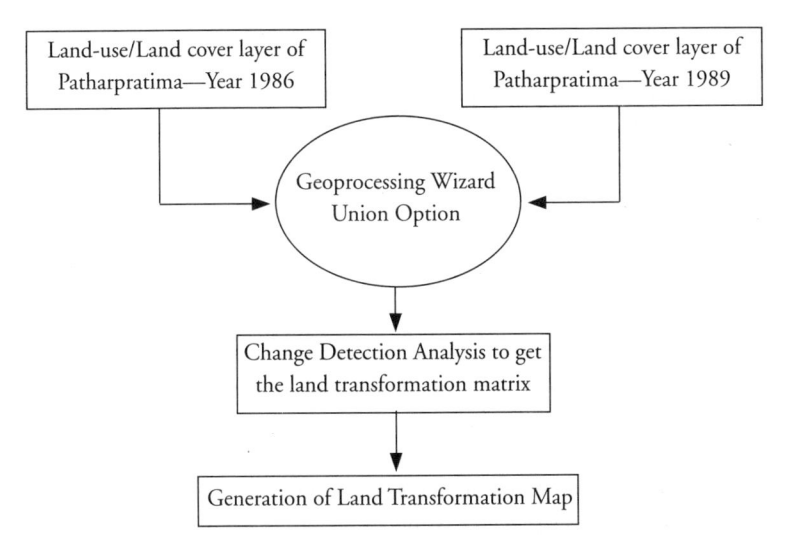

Figure 6A.4: Change Detection Procedure in ARC View Environment

A new column was added in the change detection database format of the newly generated change detection map to incorporate the changed id. Suppose a class having id 1 was changed to a class having id 2, then the new id set was given as 12.

For example, 1 was given for Dense Forest while 2 was given for Settlement with Vegetation. Then the changed id would be given as 12, the id which remained the same that is 1 to 1,2 to 2, 3 to 3 and so on, the id was given as 100 (no change), the id which was 1 and then became 0 was grouped under class-like erosion of id-200, while in case of the opposite phenomenon that is the id for a class which was 0 and after that it became 1 was grouped under class-like accretion of id-300. In this way, the new column of the change detection map was generated and the map was generated on the basis of the changed id. In this way, the same procedure was repeated for the rest of the years 1996, 2001, and 2004.

7

HUMAN WELL-BEING
WHO GAINS AND WHO LOSES?

INTRODUCTION

The impact of increasing shrimp production, processing, and export in the region on the well-being of different stakeholders in the activity is being dealt with in this chapter. Our aim is to determine the impact on levels of well-being of different groups in the region as measured by the plausible consequent changes in income, health, livelihood, and security accessible to different groups in the region. This objective in itself points towards a multi-dimensional approach to the concept of well-being and the notion that gains and losses need to be looked at in the stakeholders' perspective so as to focus on distributional issues. The aim is to determine who gains and loses and how.

Conceptual approaches to issues such as poverty, well-being and ecosystem services is being reviewed below. Subsequent sections examine stakeholder incomes as the first and most easily identifiable determinant of well-being. The next section extends the notion of well-being and sets up multi-dimensional indices of well-being for stakeholders. The chapter concludes with a comparative assessment of changes in the well-being of identified stakeholder groups. The data for the analysis in this chapter is taken from the household survey conducted as part of the survey and from secondary sources.

POVERTY, WELL-BEING AND ECOSYSTEM SERVICES:
THE LITERATURE AND SOME CONCEPTUAL APPROACHES

There exists extensive literature on poverty, its definition, measurement, and alternative conceptual approaches to it in development and environmental economics (Sen, 1999; Alikre 2007; Alkire 2002, Dasgupta 2001; Anand and Sen 2000; Osberg 2003; McGillivray 2005; Nussbaum and Sen 1993, and Qizilbash 1996). To begin with, we focus on three seminal works by Sen, Dasgupta, and Nussbaum. Sen provides a framework of welfare economics within which poverty can be studied. His 'Capability Approach' to poverty and human well-being makes a distinction between well-being, agency achievement and freedom. These concepts are related to each other but are in no way identical (Sen 1999). Here capability is defined in the space of functionings, which are related to well-being and living standards. Capability is the alternative combination of functionings feasible for an individual to achieve. Functionings, in turn, are the resources, activities, and attitudes people recognize as constitutive of well-being. Sen argues that there cannot be a canonical list of functionings; they will have to be set and reset again in different ways. In Sen's framework, capability amounts more to freedom-centric concern and issues of personal advantage (personhood) and less to achieving well-being or quality of life.

Nussbaum, on the other hand, analyses well-being from the viewpoint of ethical considerations. She maintains that 'the question whether equality of capability is or is not a good goal cannot be well answered without specifying a list of relevant capabilities.'[1] She focuses her lens on analysing the well-being through compassion. The canvas in which the well-being is seen is huge and a bit fuzzy and transcends the boundary of welfare economics and psychology (Nussbaum 2001). She suggests the study of diverse literature to examine and educate compassion and the emotions more generally. As Gasper (2003) puts it, 'Nussbaum's enormous agenda asks for many types of evidence, collaboration, and interaction'. Perception of her own degrees of empathy, compassion, mercy, and cosmopolitanism are of primary importance while defining and analysing the concept of well-being. The distinction, which exists between the approaches of Sen and Nussbaum, can be put in the form of a table:

As is clear from the above table, one can draw the inference that for a decision maker, the concept of capability is still evolving and it is neither finished nor final. Further, the multi-dimension elements of well-being also suggest that people value their freedom and

Table 7.1: Sen's and Nussbaum's Approach to Capability

Concept	An undeveloped human potential, skill, capacity	A developed human potential, skill, capacity	Set of attainable functionings given a person's skills and external conditions	A priority for attainable functioning
Sen		Capability (used in HDI etc)	Capability	Basic capability (not much in use)
Nussbaum	Basic capability, innate	Internal capability	Combined capability	Central capability

Source: Adapted from Gasper 2003.

they have reason to do so (Alikre 2007). Development is defined as *an expansion of capabilities,* leaving the selection of the relevant capabilities as a value judgement. Sen refrains from developing a list of basic capabilities or even setting up a procedure for doing so. Sen (1999) suggests:

There can be substantial debates on the particular functioning that could be included in the list of important achievements and corresponding capabilities. This valuation issue is inescapable in an evaluative exercise of this kind, and one of the main merits of the approach is the need to address these judgemental questions in an explicit way, rather than hiding them in an implicit framework.

This constitutes both a strength and a weakness of his capability approach. While it helps to avoid the problems that may emerge from an over-specification of human nature, it is open to the criticism of not being easily amenable to operationalization.[2] Development practitioners in particular see a great deal of use in drawing up lists that incorporate the different constituents and determinants of well-being. Such specifications help governments in policy direction and in measuring performance against given directions of change. They also contribute towards arresting the tendency of policy makers to couch themselves in a framework where utility (state of deriving satisfaction) is primarily and predominantly dependent upon income and subsequent consumption.

Partha Dasgupta's (2001) version of well-being is more operational and includes liberties, income, health, and education. He suggests focusing attention on the sustainability of well-being and talks in terms of a need for the comprehensive measurement of wealth. For Dasgupta, there is a thin line between well-being and quality of life and he settles down for a flexible concept of welfare where the valuation by a person of his own situation is the key. He makes an explicit preference for measuring inclusive wealth and not GNP per capita as a proxy for well-being. Here we find that his approach to well-being is arguably in a better position to capture the nuances, contribution, and the roles played by the ecosystems and the environment in human well-being.

Issues of poverty have also been discussed in the context of environment and the possible degree and direction of the relationships between ecosystem services and well-being as well (MEA 2005, Duraiappah, 1996). Poverty is defined as 'the pronounced deprivation of well-being' in the *World Development Report* (2001). The literature on well-being accepts its multi-dimensional characteristics and the *World Development Report* takes this multi-dimensionality as its reference point.

In recent years, the United Nations Development Programme (UNDP) has succeeded in operationalizing the concept of human well-being, although only in a limited sense. The Human Development Index (HDI) developed by the UNDP goes beyond income (GDP per capita) and includes health (through infant mortality rate) and education (through literacy and drop-outs from the primary school). HDI does acknowledge that the capability of the people to lead a long and healthy life, to acquire knowledge, and to have access to resources need for a decent standard of living (UNDP 2004: 127) but their index does not capture the benefits people derive from a relatively better kept ecosystem and its services like storm protection by mangroves, coast stabilization, and biodiversity maintenance by the mangrove forests such as the Sundarbans.

Simultaneously, empirical work in a large number of countries has been of use in assisting to see what values humans hold dear and in therefore identifying what may alternately be called 'basic capabilities to be striven for' or 'constituents and determinants of well being'.[3] Narayan et al. (2000b) provided a list based on evidence from a large number of developing countries. They came up with a list of constituents and determinants of well-being which was examined in a recent conceptual exercise carried out by the Millennium Ecosystem Assessment (MEA).[4] The MEA was charged with the task of examining the literature on human well-being in order to determine its links with ecosystem services. The overwhelming evidence from the studies reviewed is that poor people's idea of a good life is multi-dimensional. The dimensions cluster around the following themes: material well-being, physical well-being, social well-being, security, and freedom of choice or action. Communities scattered across different countries also included a sense of 'responsible well-being' in their understanding of what constituted the good life.[5]

Taking into account both the state of theoretical developments and empirical evidence documented, the MEA identified the constituents and determinants of human well-being as follows:

1. Basic material for a good life (adequate livelihoods, sufficient food, shelter, access to goods),
2. Health (strength, feeling well, access to fresh air, and water),
3. Social relations (social cohesion, mutual respect, ability to help others),
4. Security (Personal safety, secure resource access, security from disasters), and
5. Freedom and Choice (opportunity to be able to achieve what an individual values doing and being).

However, as has been well documented, even in the early literature, 'a major problem is that historically growth has expanded choice in some directions while constricting it in others'.[6] Such two way linkages of ecosystems, their functioning and peoples' welfare are well established and widely acknowledged in the literature (GBA 1995; Daly 1987, Dasgupta 2001, UNEP-IISD 2004, and MEA 2003

Source: MEA 2005.

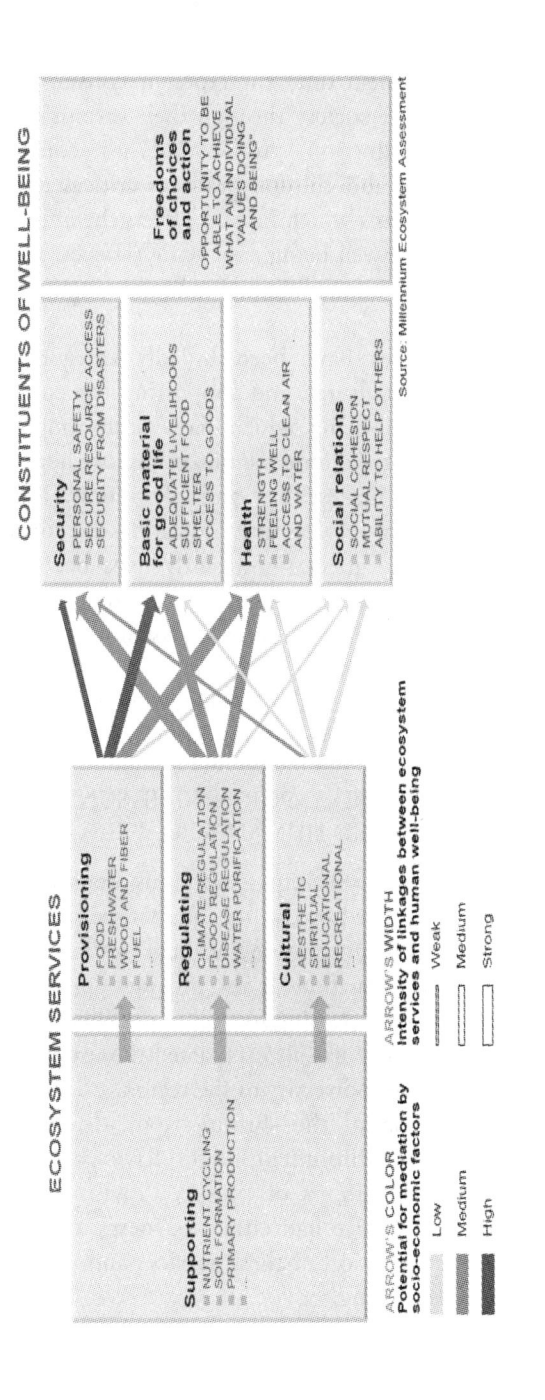

Figure 7.1: Ecosystem Services and Human Well-being

and 2005). Ecosystems yield different types of consumptive and productive benefits for the society through their various ecological functions. Society's production profile and it's consumption preference, social milieu, and cultural practices critically influence the conditions of ecosystems which in turn affects the ability of the ecosystem to deliver some well-being enhancing services. Figure 7.1 from MEA 2005 illustrates the relationships between ecosystem and human well-being.

The ecosystem services have been broadly categorized into provisioning, regulating, cultural, and supporting the constituents and determinants of well-being which have been sought to be physical security, basic material for good life, minimum health, social relations, and freedom to make choice and take action. The arrows emanating from ecosystem services and indicating the constituents of well-being which, they affect significantly have been depicted in the diagram. This schematic diagram of ecosystems and human well-being provides a perfect backdrop to analyse how the mangrove ecosystem of the Sunderbans are so central to well-being of the people. It also helps in understanding how the poverty of people can impact the shape and health of mangrove forest in the region.

SHRIMP PRODUCTION AND EXPORT IN THE SUNDARBANS, ECOSYSTEM SERVICES, AND HUMAN WELL-BEING

In the Indian Sundarbans, shrimp export has definitely earned foreign exchange. The entire process from prawn seed collection to farming, processing, and transport to the final destination creates employment and generates income. It also provides opportunities to a large number of people by creating sources of livelihood. However, as seen in earlier chapters, it has also impacted the use of land and water and influenced the biodiversity of the region.

Prior to the initiation of the shrimp export-driven changes, the natural and social environment might have provided a set of ecosystem services to groups of people. Provision of shrimp related livelihoods could have had either a synergy or a trade-off relationship with the provision of these services and goods, which also contributed to well-being.

We take here the analytical route of the impact of the shrimp export on the constituents of human well-being for the people involved. This is done by asking questions such as:

i. How does the process of export of aquaculture impact the distribution of income among different stakeholders in the region?
ii. How does land-use change in general, affect the livelihood support system of the people?
iii. Who are the losers and who are the gainers in the process of export of aquaculture from the Sundarbans?

Two types of data are used in the analysis of well-being of the stakeholders:

An estimate of income generated by shrimp farming in the Sundarbans through the activities of farming, transporting, and processing is made from secondary sources as well as data from the processing units and aquaculture farms.[7] The necessary disaggregated level data were generated through a comprehensive survey. After a detailed discussion, the pilot questionnaire was developed and subsequently adopted for the survey in the region. Survey-generated data were used together with evidence from published and unpublished government documents.

The study canvassed a detailed household questionnaire in the blocks of Minakhan, Canning, and Gosaba in the Sundarbans. Knowing the major stakeholders and the kinds of occupational choices created as a consequence of this activity, five livelihood categories were covered in the survey namely agricultural farmers, shrimp farmers, fishermen, PL collectors, and mixed income households. A multi-dimensional approach to human well-being constituted the basis for drawing up the questionnaire.[8] Indices were developed for different determinants of human well-being.

The section below discusses incomes generated in the Sundarbans as a consequence of shrimp farming. The section 'Human Well-Being for Different Stakeholders' extends the analysis to a stakeholder and multi-dimensional impacts of this income generation. All the analysis in this chapter is cross-sectional and the data relates to the year 2004.

INCOME GENERATION FROM SHRIMP EXPORTS IN THE SUNDARBANS

To begin with, incomes generated by shrimp production and export are estimated. An array of accepted methodologies for income estimation are considered appropriate. For example, the value-added approach is used for estimating incomes accruing to shrimp farmers and wage and employment data to estimate incomes of workers and PL collectors.

In 2004, the Indian Sundarbans had a total area of 42,000 hectares under aquaculture. Nearly 8,100 households were engaged in shrimp farming with an average estimated value of yield of Rs 1,35,324 per hectare. Therefore, the total estimated value of income generated for shrimp farmers stood at Rs 5.68 billion for the year 2004–5.

Since about 70 per cent of the production is exported, income generated due to export of shrimp can be put at approximately Rs 3.97 billion. Shrimp farms employ workers, buy seed from agents, in turn who get it from shrimp PL collectors. They sell their output to agents and transporters to be taken to markets or to processing units.

Table 7.2: Block-wise Average Income of a Shrimp Farmer (Rs/Annum)

Block	Per Household	Per Capita	Per Hectare (Gross Value)	Per Hectare (Net Value)
Minakhan	31,455	7,302.43	1,30,036.3	70,710.6
Canning I&II	38,940	9,350.67	2,97,826.3	1,99,938.9
Gosaba (culture ponds)	28,164	5,522.97	–	–
Average	21,153	7,392.02	2,13,931.3	1,35,324.7

Source: Authors' calculations.
Note: Column 5 shows per hectare value for the year 2004 but not annual value. In Gosaba, most of the farms are only PL culturing units and, therefore, are not treated as shrimp farms. They are, however, a part of the production chain and generate incomes.

Table 7.3: Block-wise Average Income of a Shrimp Farm Worker

(Rs/Annum)

Block	Per Household	Per Capita
Minakhan	24,360	5,228
Canning I & II	19,200	3,840
Gosaba	–	–

Source: Authors' calculations.

Shrimp Farm Workers: There are about 72,000 shrimp farm workers in the Indian Sundarbans with an estimated average annual wage income of Rs 18,000 per worker.[9] Therefore, the total annual income accruing to the shrimp farm workers in the Sundarbans area is about Rs 1.29 billion.

Shrimp PL Collectors: There are around 0.15 million shrimp *post larvae* (PL) collectors in the Indian Sundarbans.[10] Their average annual per capita income is approximately Rs 5,000. The total estimated income generated by these shrimp PL collectors is about Rs 0.75 billion for the year 2004–5.

Processing Units: The processing units in Kolkata employ more than 17,000 workers and generate an estimated annual income of Rs 0.04 billion for workers. The units' profit could be between 2 to 5 per cent of their turnover, amounting to Rs 0.10 to Rs 0.25 billion.

Table 7.4: Block-wise Average Income of a PL Collector

(Rs/Annum)

Block	Per Household	Per Capita
Minakhan	–	–
Canning I & II	22,280.00	4,986.00
Gosaba	26,284.20	5,020.58

Source: Authors' calculations.
Note: No PL collector reported from Minakhan in the survey.

Table 7.5: Income and Employment from Shrimp Production and Processing for Export in the Indian Sundarbans
(Annual)

Stakeholder	Number ('000)	Income (Rs Billion)	Attributable to Export Activity (Rs Billion)
Shrimp Farmers	8,100	5.68 (7.79)*	3.98
PL Collectors	150	0.75	0.52
Shrimp Farm Workers	72.14	1.29	0.90
Processing Unit value added	–	0.24	0.24
Transporters and Agents	–	0.90	0.90
Total		8.87	6.55

Source: Authors' calculations
Note: All figures are annual for the year 2004–5.
* Figures in brackets under category shrimp farmers are estimated on the basis of land-use data from NRSA.
Note: No shrimp farm worker reported from Gosaba in the survey.

Transport and Trade Margins: These are estimated on the basis of the difference in price of shrimp at the farm gate and as purchased by the processing unit. This is Rs 65 per kg. on an average and implies that the transport and retail margins generate an income of Rs 0.90 billion.

A comparison of incomes accruing to different agents is in order. A shrimp farm owner gets the maximum per capita income as compared to prawn seed collectors, farm workers, and workers in the processing unit. This is true for all the blocks—Minakhan, Gosaba, and Canning. However, an agricultural farm household has an income of Rs 17,324 compared to Rs 32,853 obtained by a shrimp-farming household. It is, therefore, easy to understand the motivation for land-use conversion from agriculture to aquaculture.

PL collectors also have a household income of Rs 24, 282, with very little investment of capital. Given the very good market

Table 7.6: Block-wise Average Income of an Agricultural Farmer
(Rs/Annum)

Block	Per Household	Per Capita	Per Hectare (Net Value)
Minakhan	18,000	4,290.00	3,989.63
Canning I&II	13,410	2,485.00	1,038.43
Gosaba	20,562	3,348.30	1,038.43

Source: Authors' calculations.
Note: Column 3 shows per hectare value for the year 2004 but not annual value.

linkages and the minimal investment, which too is made, by the aratdar or agent, it would seem that they are getting this income almost entirely due to their effort or labour and the abundance of the natural resources rich region in which they live. In fact, they have a higher income than agriculture farm households and shrimp farm workers in the same region. Are they better off by other indices of well-being as well? Would they choose this occupation if they had other options open to them? The next section tries to answer some of these questions using an array of well-being indices.

HUMAN WELL-BEING INDICES FOR DIFFERENT STAKEHOLDERS

The household survey conducted in Canning, Minakhan, and Gosaba, covering more than 150 households provided the information necessary to estimate the human well-being indices in such a highly differentiated manner. Table 7.7 gives the manner in which the indices are constructed from the information collected.

An index of income security, for instance, is constructed in different ways depending on the stakeholder group involved. Life security for instance is a hazard faced only for fishermen and PL collectors, who venture into the water, and more so for fishermen. Chronic exposure to a form of health hazard through standing in brackish and salt water may be an important source of ill-being for PL collectors.

To further elucidate whether the five indices selected are similar in a qualitative sense, the manner in which they are operationalized

Table 7.7: Measurement of Indices of Well-being/Ill-being

Indicators	Measured by
Income Per Capita	Based on 0 to 1 scale. Rs 10,000 per annum representing 1.
Income/Food Security	*For agricultural farmers* Based on 0 to 1 scale, depending on percentage of area lost to flooding
	For shrimp farmers: Based on 0 to 1 scale, the number of normal harvests in the last five years. 0 being no normal harvest in the last five years.
	For wage-salary earners and mixed income households: Probability of getting work in a month. With 1 being maximum probability.
	For fishermen and PL collectors: Percentage who collects fuel wood and honey along with fish for self-consumption and sale in the market adds.
	For PL collectors: Probability of number of days of PL collection in a month. With 1 being the maximum probability.
Health/ Life Security	*For all stakeholders*: Based on 0 to 1 scale, inverse of Health Expenditure as a percentage of total expenditure. Often calculated as based on number of working days lost or chronic ailments suffered.
	For fishermen and PL collectors: Based on 0 to 1 scale, inverse of the average of number of storms and accidents in the last five years. With 0 being no storms and accidents and 1 being the maximum number of storms and accidents. Lack of adequate weather information lead to these.
Absence of conflict/ Social Cohesion	Inverse of conflict, which is based on 0 to 1 scale, 1 representing maximum social cohesion.
	Nature of conflict varies: for farmers: for fisherman conflict over fishing rights; for PL collectors from C.M. Khadi (a village surveyed), it was conflict with traders over price of PL; Also conflicts over forest entry for collection, conflict over title to land in the case of shrimp farmers

Source: Authors' definitions.

takes into account the space within which diverse stakeholders 'function'. Income or consumption security is interpreted as the probability of getting work for wage earners. For other households, it is approximated by the probability of PL collection days or percentage of households collecting honey and fish along with other household tasks.

While social cohesion or absence of conflict is a component in the well-being of all, it is captured through diverse manners for different groups. For shrimp farmers, it is measured through absence of conflict over titles to land; for PL collectors on the other hand, absence of conflict with traders on price is important.

HUMAN WELL-BEING INDICES

In interpreting the results and comparing well-being levels, spatial differences play a role too. Canning and Minakhan are the blocks with a greater concentration of shrimp farms, where different technologies for shrimp farming have been adopted and a move towards polyculture with improved traditional techniques found to be the best option. These are also the blocks where land conversion has been from agriculture to aquaculture and then to brick kilns. Gosaba is the area with larger concentration of fishermen, PL collectors, and where alternative livelihood experiments have also emerged. It includes the relatively distant and less accessible region as well.

For the study across regions and stakeholders, we have selected the following sets of stakeholders for comparison:

Shrimp farmers and agricultural farmers in Canning and Minakhan: This comparison is significant in view of the land-use change from agriculture to aquaculture in the region in the period 1986 to 2004. In other words, the query addressed is: what have been the human well-being impacts of this change.

Shrimp farmers and mixed income households in Canning and Minakhan: This comparison assumes significance in view of the fact that it represents the option that might have existed for upwardly mobile agricultural farmers in case the export-based expansion of aquaculture had not taken place.

PL collectors and fisherman in Gosaba: PL collection has emerged as a major livelihood source in recent years and it is mainly fisherfolk who have digressed into this. The consequence is a change in the manner of use of the natural resource base, at a cost of biodiversity loss. How has this made the PL collectors better off in comparison with fishermen?

PL collectors and salary and wage earners' households in Gosaba. Salary-wage earners' households represent the options available to households as a consequence of better educational levels in comparison to PL collectors. They also illustrate the possible alternative livelihood strategies in the region, if PL collection is to be banned. It is worthwhile to see how this comparison performs.

SHRIMP FARMERS AND AGRICULTURAL FARMERS IN CANNING AND MINAKHAN

As indicated in Tables 7.8 and 7.9, shrimp farmers in Canning and Minakhan are better off than agriculturists in terms of per capita incomes with indices in the range of 0.56 and 0.93 as

Table 7.8: Indicators of Well-being of Agricultural Farmers in Canning and Minakhan

Indicators	Canning	Minakhan	Measured by
Income Per Capita	0.25	0.43	Based on 0 to 1 scale. Rs 10,000 per annum representing 1.
Income Security	1.00	1.00	Based on 0 to 1 scale, inverse of percentage of net sown area not cultivable due to flooding.
Health	0.91	0.84	Based on 0 to 1 scale, inverse of Health Expenditure as a percentage of total expenditure.
Absence of Conflict/ Social Cohesion	1.00	1.00	Inverse of conflict, which is based on 0 to 1 scale, 1 representing maximum number of conflicts.

Source: Authors' calculations.

Table 7.9: Indicators of Well-being of Shrimp Farmers in Canning and Minakhan

Indicators	Canning	Minakhan	Measured by
Income Per Capita	0.93	0.73	Based on 0 to 1 scale. Rs 10,000 per annum representing 1.
Income Security	0.66	0.56	Based on 0 to 1 scale, the number of normal harvests in the last five years. 0 being no normal harvest in the last five years.
Health	0.94	0.94	Based on 0 to 1 scale, inverse of Health Expenditure as a percentage of total expenditure.
Absence of Conflict/ Social Cohesion	0.50	0.50	Inverse of percentage of farmers reporting farm related disputes.

Source: Authors' calculations.

compared to indices between 0.25 and 0.43. However, the security of these incomes is much lower in terms of frequency of normal harvests. Also, while agricultural farmers are not exposed to any conflict, shrimp farmers have a relatively lower level of social cohesion due to constant conflicts. A large number of these conflicts are over property rights to land for ponds. In terms of a spatial comparison, shrimp farmers in Minakhan seem to be worse off than those in Canning. For agriculturists, it is not so easy to rank the two blocks. Figure 7.2 shows the same indices in graphic form, using a 'spider diagram'.

In this genre of diagrams, distances from the centre, along the four axes, measure indices for different components of well-being. Each quadrilateral stands for the mix of indices for stakeholder categories. The differing situations for stakeholder categories can be read with ease from the diagrams.

Figure showing radar chart with axes: Income Per Capita, Income Security, Social Cohesion, Health Security. Legend indicates Shrimp Farmers Canning, Shrimp Farmers Minakhan, Agricultural Farmers Canning, Agricultural Farmers Minakhan, with arrows for Decreasing and Increasing.

Source: Authors' calculations.

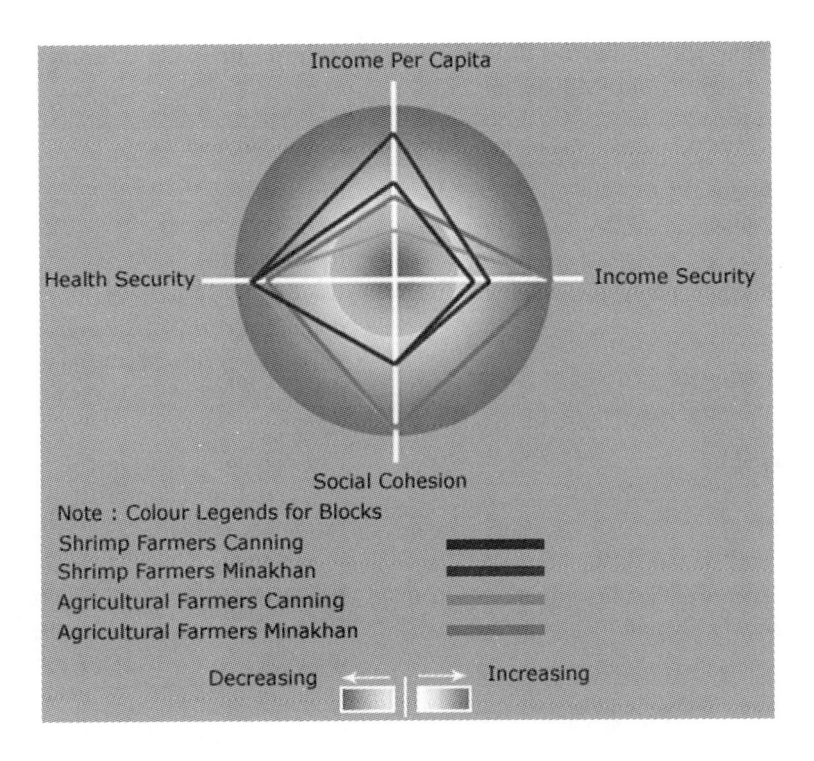

Figure 7.2: Indicators of Well-being of Shrimp Farmers and Agricultural Farmers in Canning and Minakhan

SHRIMP FARMERS AND MIXED INCOME HOUSEHOLDS IN CANNING AND MINAKHAN

Mixed Income households in Canning and Minakhan have lower levels of income compared to shrimp farmers but enjoy more security with respect to availability of this income. They have lower levels of conflict since they do not have to face land-related property rights issues.

Spatial comparisons are also interesting. Shrimp farmers in Minakhan are worse off than shrimp farmers in Canning on all indices. However, mixed income households in Minakhan are better off than mixed income households in Canning. It could be that Canning has

Table 7.10: Indicators of Well-being of Shrimp Farmers and Mixed Income Households in Canning and Minakhan

Indicators	Mixed Income Households Canning	Mixed Income Households Minakhan	Shrimp Farmers Canning	Shrimp Farmers Minakhan	Measured by
Income per capita	0.56	0.58	0.93	0.73	Based on 0 to 1 scale. Rs 10,000 representing 1.
Income security	0.70	0.75	0.66	0.56	Probability of getting work in a month with 1 being maximum probability.
Health	0.92	0.91	0.94	0.94	Based on 0 to 1 scale, inverse of health expenditure as a percentage of total expenditure.
Absence of conflict/ social cohesion	0.9	1.00	0.5	0.5	Inverse of conflict, which is based on 0 to 1 scale, 1 representing maximum number of conflicts.

Source: Authors' calculations.

a comparative advantage in shrimp farming whereas Minakhan has the same with respect to income from other activities.

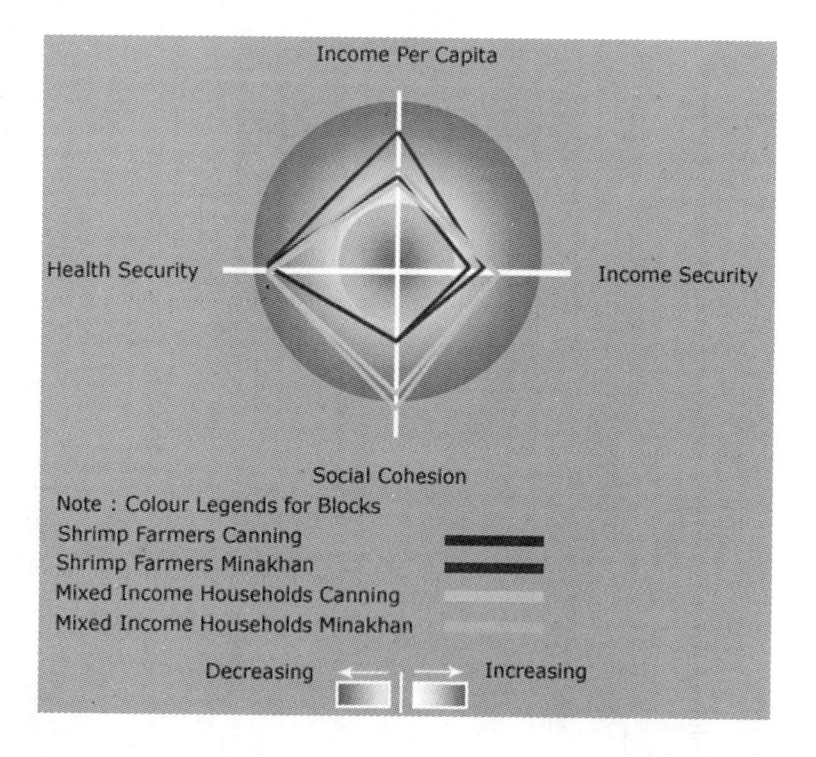

Source: Authors' calculations.

Figure 7.3: Indicators of Well-being of Shrimp Farmers and Mixed Income Households in Canning and Minakhan

PL Collectors and Fishermen in Gosaba

In Gosaba, the increase in shrimp farming has led to the emergence and consolidation of PL collection as a source of livelihood. When compared to fishermen, this has meant an increase in income and a decreased risk to life, since they need not venture into the high seas. Food security from collection of food from water bodies remains unaffected. Additionally, there is a higher level of income, food, and life security. This comparative gain and loss is illustrated in Figure 7.4.

However, chronic health problems arising out of the nature of the work increased. 85 per cent complained about these problems,

Table 7.11: Indicators of Well-being of PL Collectors and Fishermen in Gosaba

Indicators	PL Collectors Gosaba	Fishermen Gosaba	Measured by
Income per capita	0.50	0.30	Based on 0 to 1 scale. Rs 10,000 per annum representing 1.
Income security	0.40	0.60	Based on probability of number of days of collection of fish/PL in the month.
Food security	0.65	0.60	Percentage of fishermen/ PL collectors who collect fuel, wood, and honey along with fish for self-consumption and sale in the market.
Life security	0.90	0.60	Based on 0 to 1 scale, inverse of the average of number of storms and accidents in the last five years. (With 0 being no storms and accidents and 1 being the maximum number of storms and accidents.)
Health	0.38	0.36	Based on 0 to 1 scale, inverse of health expenditure as a percentage of total expenditure/ or complaints regarding chronic health ailments
Absence of conflict/ social cohesion	0.50	0.75	Inverse of conflict, which is based on 0 to 1 scale, (1 representing maximum number of conflicts.)

Source: Authors' calculations.

though they did not incur much expenditure on them. Conflict increased on several counts: 75 per cent of PL collectors reported conflicts over price with the trader. 55 per cent expressed dependence on the dadun for sale of catch. The same person also lent them money for buying the net. About 87 per cent received this credit facility. Their options with regard to sale were then limited.

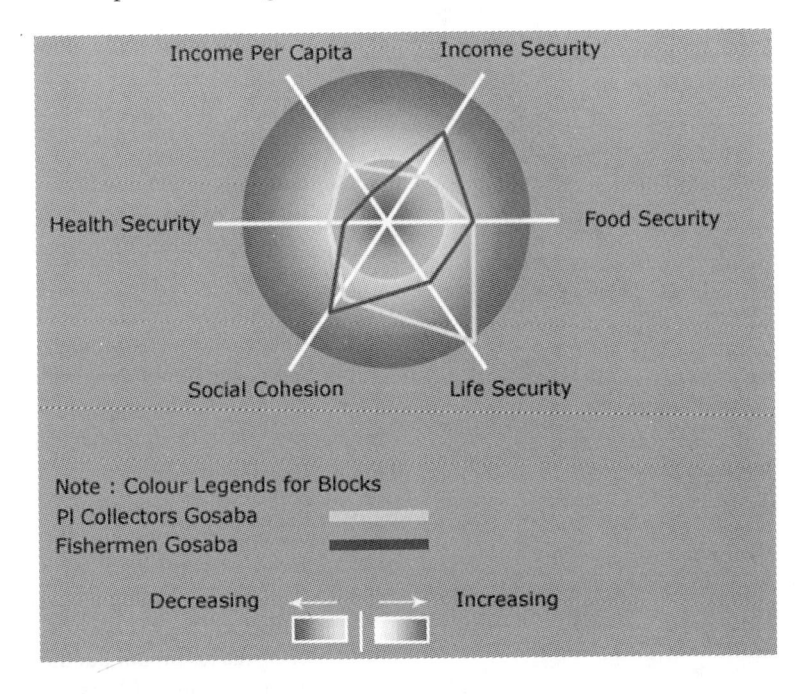

Source: Authors' work.

Figure 7.4: Indicators of Well-being of PL Collectors and Fishermen in Gosaba

PL COLLECTORS AND SALARY-WAGE EARNERS IN GOSABA

When compared to PL collectors, salary-wage earner households have higher income, more security of income, lesser health problems, and minimal conflicts. Conflicts they recall deal with law and order problems. The salary-wage earner households have higher levels of literacy with at least half the family literate. They travel to the mainland for work. On all indices, salary and wage earners are better

Table 7.12: Indicators of Well-being of PL Collectors and Salary/Wage Earner in Gosaba.

Indicators	PL Collectors Gosaba	Salary/Wage Earner Gosaba	Measured by
Income per capita	0.50	0.94	Based on 0 to 1 scale. Rs 10,000 per annum representing 1.
Income security	0.40	1.0	Probability of getting work in a month. With 1 being maximum probability.
Health	0.38	0.90	Based on 0 to 1 scale, Inverse of Health Expenditure as a percentage of total expenditure.
Absence of conflict/ social cohesion	0.50	1.0	Inverse of conflict, which is based on 0 to 1 scale, 1 representing maximum number of conflicts.

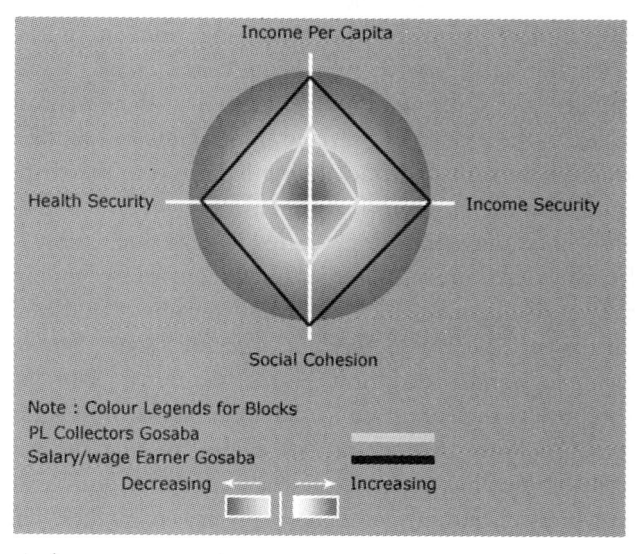

Source: Authors' calculations.

Source: Authors.

Figure 7.5: Indicators of Well-being of PL Collectors and Salary/Wage Earner in Gosaba

off than PL collectors. This has important implications for livelihood options, which can be promoted, in the region.

CONCLUDING REMARKS: INCOME GENERATION, WELL-BEING LEVELS, AND RESOURCE USE IN SHRIMP CULTURE

In interpreting incomes generated due to shrimp-related economic activity and well-being indices of different stakeholders, one finds that incomes are indeed generated in the region. In the year 2003–4, an export activity worth about Rs 5.08 billion at international prices resulted in income generation of Rs 6.55 billion. Every rupee of export led to regional income accrual of Rs 1.29.

The distribution of income generated was as follows: shrimp farmers 60.94 per cent, shrimp farm workers, 13.77 per cent, PL collectors, 8.01 per cent, transporters and retailers: 13.81 per cent, and processing units with 3.71 per cent.

However, the following types of instability characterized this process of income generation:

1. Unpredictable and increasingly complex international environment faced by the processing units in a situation where non-tariff barriers were being characterized by increased stringency. Exporting units claimed that they were able to adjust to these changes provided information on standards was available on time and traceability of shrimp to the farmer was ensured to ensure appropriate practices.
2. Shrimp-farming technology related income fluctuations. In the late-nineties, farmers observed that intensive technology was subject to unpredictable disease and incomes could be highly fluctuating. They also faced a Supreme Court ban on this technology and settled for improved traditional technology.
3. Land-use change occurred in the main blocks from agriculture to aquaculture. In some blocks, it also meant transformation of mangroves into aquaculture farms. High margins associated with shrimp farming were the main drivers of these changes, and
4. Absence of hatcheries and their high cost meant that PL collection from the wild was the major basis of shrimp culture. This led to a declining trend in biodiversity indices in high collection zones.

This imposed a social cost, which could have been internalized within the cost of shrimp farming given the high market prices of shrimp. However, due to the absence of institutions for doing the same, it was not done.

Human well-being indices based on a multi-dimensional approach to well-being provides an insight into changes brought about in the lives of different stakeholders:

1. Shrimp farmers in Canning and Minakhan are better off than agriculturists in terms of per capita incomes. However, the security of these incomes is much lower in terms of frequency of normal harvests. Also, while agricultural farmers are not exposed to any conflict, shrimp farmers have a relatively lower level of social cohesion due to constant conflicts. Aquaculture farmers had to face higher conflict situations and more income insecurity than agricultural farmers.
2. PL collectors have increased incomes and reduced life insecurity, when compared to fishermen. They do not enjoy the level of income security, health security, and conflict free livelihoods as salary and wage earning households in the same area have. Alternative livelihoods based on the latter seem a viable option.

Based on the above and other supporting evidence collected, it would be safe to conclude that:

1. Sundarbans is witnessing environmental changes in various forms such as PL collection activity leading to soil erosion, thereby depleting the mangroves, and incurring heavy biodiversity losses in terms of killing many finfish varieties. PL collection activity is becoming not only difficult physically but economically too. Since the supply of the wild PL is not unlimited and continuous, therefore, the PL collectors have to spend more time and cover longer distances in search of the wild PL. On the other hand, the demand from the shrimp-processing firms is increasing overtime.
2. Similarly, the aquaculture activity also has negative consequences on the environment in the region. The aquaculture ponds/farms require huge amount of land and water resources. The land for

this activity comes either through conversions from paddy fields or mangroves. Once the land is converted for aquaculture, chances of its reverting to other kinds of agriculture are fewer, though there are some evidences of paddy-cum-aquaculture as well.

3. Most of the aquaculture ponds in the study region employ traditional farming techniques and are maintained properly throughout the culture period. So, they are sustainable except during the times of some serious disease outbreaks. Farmers in Canning blamed the viral disease problem to the bad feed imported from Thailand under the World Bank project in that area during 1998–2000. Inspite of separate water supply and drainage canals, an outbreak of disease in any of the farms affects almost all the farms.

4. Though, there are no major water pollution problems caused by the aquaculture activity in the region, settling of fly ash of some neighbouring brick kiln units on the water surface of the farms in some areas has caused problems for them. The farmers blame the kilns for emitting poisonous smoke (SPM) which settles on the surface water of their farms, thereby causing suffocation to the fish and shrimp population. In fact, the visibly thick SPM layers hamper the process of photosynthesis, resulting in a reduction of the planktonic growth in these farms.

5. These brick kiln units are licensed by government and adhere to the environmental norms—such as height of chimneys, distance from river etc. But the main concern of the local farmers is that such norms do not take care of the aquaculture farms since no feasibility study was carried out to see the impact of brick kilns' operation on the health of various fish varieties. The emission of the ash and SPM from the chimneys has increased the cost of treatment of farms manifold and so the profit margins have dropped.

6. Another aspect of the brick kiln business is that once these units close down, the areas cannot be used either for shrimp production or paddy cultivation in future, which is an economic and ecological loss.

7. Definitely, any reductions in export opportunities or fall in profits would have negative impact on the overall shrimp production

activity. Though, marine products have quite high domestic demand in the region, but the scale at which the shrimp farms and processing units are operating presently, is based, primarily, on the lucrative export business. Once the export opportunities are exhausted or weakened due to any measure/reason, there will be a sharp reduction in the overall shrimp production.

8. PL collection is worse than fishing for biodiversity at any given point of time since this activity erodes soil and mangrove saplings at the river fronts and involves huge loss of other finfish varieties. On the other hand, fishing is undertaken, mainly in deep waters and without much loss of fish and mangrove biodiversity.

9. Out of the category of wagers/salaried people, the latter mainly constitutes people engaged permanently in agricultural fields and shrimp farms; teaching, and government offices. These options require either some formal education or physically strenuous work, which is not possible for the illiterate, poor, and mainly women PL collectors. So, there is hardly any chance of these people moving to the salaried class.

NOTES

1. See Chapter 3, by Nussbaum in Agarwal, Humphries, and Robeyrs, edited (2006).
2. Sugden (1993) gives one such critique. Stewart (1996), Gasper (1996) are others who advocate movement towards a basic set of capabilities. Nussbaum (2000) has in fact, drawn up a list of central human capabilities.
3. See Dasgupta (2001) in this context.
4. See Millennium Ecosystem Assessment (2003), in particular Chapter 3 on Human well-being.
5. This led Chambers (1997) to consider 'responsible well-being as the main agenda for development'.
6. See Appffel-Marglin and Marglin (1990).
7. See Appendix 7A for the methodology used in this estimation.
8. See Appendix at the end for the questionnaire.
9. This is estimated based on the prevailing wage rate and the number of days of work.
10. This figure is from latest survey carried out by Abhijit Mitra (2005).

Appendix 7A

METHODOLOGY FOR ESTIMATING INCOMES GENERATED BY AQUACULTURE IN THE INDIAN SUNDARBANS

7A.1 SHRIMP FARMERS

(i) Number of Shrimp Farmers = Land under Aquaculture (ha) / Avg. land holding per farmer (ha)
  - Land under aquaculture—Different studies have reported different figures for the land under aquaculture in Sundarbans. Some of them are as follows:

Table 7A1: Figures for land under aquaculture in Sundarbans

Study	Year	Area under Aquaculture (Ha)
Govt. of West Bengal, A draft report on Sundarban Wetland	2000	45,000
Indian Hatcheries Organization (1999) as reported in L. Hein, 'Impact of Shrimp Farming on mangroves along India's East Coast'	2002	45,525
National Biodiversity Strategy Action Plan Report, Tamil Nadu	2003	42,750
Review of Fisheries Sector of Sundarbans	2004	42,000
School of Oceanographic Studies, JDU, Kolkata (based on NRSA data for 1986–2004)	2004	57,585

Source: NRSA (1986–2004).
Note: This variation in data leads us to think as to which value should we take for our estimation of shrimp farmers and their income generated. We took a figure of 42,000 hectares.

  - *Average Land Holding (Ha)*—Data on average land holding per farmer (ha) is taken from the primary survey of the aquaculture carried out in Canning-I and Minakhan blocks.

(ii) *Income of Shrimp Farmers*—Total area under aquaculture (ha)* Average net value of yield from aquaculture (Rs/ha)

7A.2 SHRIMP PL COLLECTORS

(i) Income of Shrimp PL collectors = Number of shrimp PL collectors* Average per capita income (Rs/annum)
- Data on number of shrimp PL collectors is stated as 4 lakh in the Central Inland Fisheries Research Institute of India (CIFRI) estimates. The latest survey from Mitra of the Department of Marine Sciences gives a smaller figure of 1.5 lakhs. We use the later figure.
- Data on an average per capita income is used from the primary survey.

7A.3 SHRIMP FARM WORKERS

(i) Number of shrimp farm workers[1]—10 per cent of the total agricultural workers (main and marginal) in Sundarbans
(ii) Income of shrimp farm workers—number of shrimp farm workers*average annual income per worker (Rs/annum)

7A.4 PROCESSING UNITS' MARGINS AND WORKERS WAGES

The margins or net profit of the processing unit is taken to be from 2 to 5 per cent of the value of their gross turnover. Workers' wages are found thus:

(i) Number of workers = Sum of all skilled, semi-skilled, and unskilled workers, and
(ii) Income of workers = Sum of incomes for all categories of workers.

Note: Both these data are taken from the primary survey of processing units in Kolkata.

7A.5 TRADE AND TRANSPORT MARGIN

These margins per unit of shrimp processed are calculated from the difference between the per unit farm gate price of shrimp and the price paid by the processing units. When multiplied by amount of shrimp processed for export (this figure of Rs 65 per kg), they give the income of transporters and retailers.

Appendix 7B

THE CHOTTO MOLLAKHALI (GOSABA BLOCK) SURVEY[2]

A household survey of 100 households was conducted in Chotto Mollakhali Gram Panchayat, Block Gosaba, District South 24 Parganas in West Bengal. For each of the five livelihood categories, 20 households were selected at random. Though randomly selected, survey locations were predetermined where there was high probability of finding the target households. As per the 2001 Census, Chotto Mollakhali has a population of 18,419 of which 65 per cent are Scheduled Castes (SCs). However, the sample has a disproportionately high percentage of SCs; just three families are non-SCs. This may be attributed to the fact that localities like Adivasipara and Chotto Mollakahli proper have not been covered for surveying the target groups except for the salaried class. The Agricultural Farmer Household Survey was carried out in Mistrypara, Chotto Mollakhali village; Shrimp Farmer Household Survey was carried out in Gobindapur village and 9 no. Para (neighbourhood or locality), Kalidaspur village since there are two different kinds of farmers at the two locations. Fisherman Household Survey was carried out in Para Kalidaspur village, Shrimp PL Collector Household Survey was carried out in Gobindapur and Marichjhapi, which are essentially adjoining localities. The Wage/Salary Earner Household Survey was carried out in Chhoto Mollakhali proper, since there is a concentration of salaried class in the locality. Like many other parts of Sundarbans but unlike Sagar and Namkhana blocks, Chhoto Mollakhali does not have many households whose only or primary means of livelihood is dependent on daily wage. In Chhoto Mollakhali, most families have to depend on a package of livelihood options and therefore, the survey was carried out amongst households whose primary means of livelihood fitted with one of the target categories. Accordingly, wage/salary earner household survey was restricted to the employed class.

In the Chhoto Mollakhali area of shrimp farming, ponds varied between less than 2 bighas to more than 20 bighas. 55 per cent of the households have taken to shrimp farming during the last five years or less, 35 per cent during the past 10 years or less, and only 10 per cent have been in business for about 20 years.

Though only 20 per cent of the farmers reported normal to good harvest in the last five years and 50 per cent for only two years, 85 per cent of the respondents did not want to change their profession. *One reason for this disinclination could be that 75 per cent of the farmers have secondary sources of earning.*

COLLECTION OF ITEMS BY SHRIMP FARMER HOUSEHOLDS

Four families (20 per cent) collect multiple items and 11 families (55 per cent) collect a single item, be it from land or water. Items collected are fish from rivers and creeks, firewood from government forest,[3] and water for irrigation from private as well as common property. Men collect all the items. Fish is meant for regular sale to local traders for prices ranging between Rs 30 and 50. Local traders do not extend any credit facility. Firewood is for self-consumption. Eight respondents (40 per cent) encountered problems while collecting items from land and water, one (5 per cent) reported conflict with others and the rest (55 per cent) complained of health problems. Respondents had problems quantifying man-days lost or cost incurred in solving health problems. Fishing trips last for seven days or more and it is not infrequent that fishermen feel ill during this period but continue fishing.

The average monthly household expenditure varies between one thousand five hundred rupees or less to over three thousand rupees. The distribution is as under:

Table 7B.1: Average Monthly Household Expenditure of Shrimp Farmers

Average Expenditure (Rs)	≤ 1500	≤ 2000	> 2000	> 3000
No. of Households	3 (15%)	8 (40%)	5 (25%)	4 (20%)

Source: Centre for Environment and Development, 2005.

Principally, the drinking water source is the tubewell for all the households. For lighting, 19 families (95 per cent) use kerosene and one family (5 per cent) uses solar power. Wood is the primary cooking fuel supplemented by dung and straw.

40 per cent of the households do not own any of the listed household items while 12 own bicycles (60 per cent), five of these families also own a boat (25 per cent), and one respondent had a TV and telephone (5 per cent). None of the households buy any newspaper; they listen to radio but not necessarily the news.

Eight of the families (40 per cent) have borrowed money for shrimp farming from private lenders, five of whom (25 per cent) have borrowed up to 10,000 rupees. One (5 per cent) has borrowed Rs 50,000 and another (5 per cent) Rs 100,000. *Only one family recalled a dispute over drinking water in the locality.*

All fishermen complained about the lack of adequate weather information at fishing sites[4] and recalled encountering heavy storms at least twice during the past five years, 15 per cent reported at least six encounters. About 55 per cent of the fishermen have heard of at least nine accidents (possibly death) at fishing sites during the past five years. However, due to lack of clarity of the term 'accident', one fisherman said he had heard of about 100 accidents (possibly not deaths) during the 5-year period. Seventy per cent of the fishermen complained of health problems.

Price demanded by fishermen and price offered by sales agent/ trader often leads to quarrels, 75 per cent reported problems regarding selling their catch. This is to be expected since there are no sales cooperative in the area.

Conflicts related to fishing rights are common, and 50 per cent of the fishermen recalled recent conflicts but at the same time refused to change their profession except one who preferred shifting to agriculture.

SHRIMP PL COLLECTOR HOUSEHOLDS

All the respondent families are Scheduled Castes who own their dwellings, all of which are *kutcha* structures. Shrimp PL collector families are generally large; 30 per cent households reported family size of more than four members but less than six, and 45 per cent reported family size greater than or equal to six. Fourteen (70 per cent) of the families were found to have more than half the family members literate. Amongst the 20 households, there are 39 earning members, five of whom are females; and twenty of the earning members travel outside the island for work. About 75 per cent of the families have two or more earning members. Thirteen (65 per cent) families reported at least one member who contributes through either fishing or collection of forest produce.

As with the fishermen households, respondents of shrimp PL collector families were unable to distinguish between river and creek, and all reported PL collection in river. Eleven (55 per cent) households spend about 15 days collecting shrimp PL; six (30 per cent) households spend about 7 days, and the rest about 20 days. Earning per day varied between Rs 20 and Rs 50; 35 per cent reported their earning to be around Rs 20, another 35 per cent put it at Rs 50, and for the rest 30 per cent it varied between Rs 30 and 40 a day. Eighty per cent households reported PL collection for about 5–6 months; the rest said they collected PL for about nine months.

Seventy per cent of the households surveyed reported buying their nets during the past 2–3 years for a price mostly varying between Rs 500 or less (50 per cent) to Rs 1500 or less (25 per cent); for the rest it was between Rs 3,000 and 10,000. About 45 per cent of the households financed their own nets, 40 per cent depended on money lenders, while the rest bought through borrowing from the sales agent / trader. The nets are relatively old, 75 per cent expect the nets to last for just another year.

Per day average collection generally varies between 40 and 100 for majority (75 per cent) of the households. All the households sell it to the trader (Dadun) for a price ranging between Rs 200 to Rs 400 per 1,000.

Sixty five per cent complained about lack of adequate weather information at collection sites, and 80 per cent recalled between two and ten accidents during the past five years. All the collectors complained of health problems.

Price demanded by collectors and price offered by traders often leads to quarrels, and 75 per cent reported problems regarding selling their catch. This is to be expected since there are no sales cooperatives in the area.

COLLECTION OF ITEMS BY SHRIMP PL COLLECTOR HOUSEHOLDS

Three families (15 per cent) collect single items and 17 families (85 per cent) collect multiple items, be it from land or water. Items collected are honey and firewood from government forest, and fish from rivers and creeks. Men collect all the items. Seventy per cent sell firewood on a regular basis for Rs 1.50 to Rs 2 per kg,[5] and honey is sold by 75 per cent of the households mostly to traders for Rs 30–50 per kg. Fish is mostly (about 85 per cent) for self-consumption. The village trader is preferred for selling the collected items from whom about 87 per cent of the sellers receive credit facility. Three respondents (15 per cent) did not encounter any problem while collecting items from land and water, the rest complained of health problems (35 per cent), trouble with the Forest Department (55 per cent), and also conflict with other groups (5 per cent). Respondents had problems quantifying man-days lost or cost incurred in solving health problems. All households that run into trouble with the Forest Department for illegal entry into the forests are able to quantify cost in terms of man-days lost as well as money spent to solve the problem.

The average monthly household expenditure varies between one thousand five hundred rupees or less to over three thousand rupees. The distribution is as under:

Table 7B.2: Average Monthly Household Expenditure of
PL Collectors

Average Expenditure (Rs)	≤ 1500	≤ 2000	> 2000	> 3000
No. of Households	3 (15 %)	6 (30 %)	8 (40%)	3 (15%)

Source: Centre for Environment and Development, 2005.

Principally, drinking water source is the tubewell for all the households. For lighting, all the families use kerosene. Firewood is the main cooking fuel supplemented by dung and straw by four households.

All the 20 families own boats, while 7 (35 per cent) own bicycles as well; one family also has a TV set. None of the households buy any newspaper; they listen to radio but not necessarily the news.

Fifteen of the families (75 per cent) have borrowed money; 55 per cent for fishing equipment, and one household each for debt repayment, agriculture, shrimp farming, and 'other' uses. Only two households have borrowed from the bank, the rest either from the private lender (55 per cent) or from the family (5 per cent), or 'other' (5 per cent). Most (80 per cent) of the borrowings are between Rs 1,000 and Rs 3,000, only two households (13 per cent) borrowed Rs 10,000 or less.

Eleven families (55 per cent) recalled a problem in the area; mostly (82 per cent) over forest entry, and 9 per cent each over sale / trade and drinking water.

NOTES (APPENDIX)

1. Alternatively, number of shrimp farm workers could be estimated by multiplying the average number of shrimp farm workers (estimated from our survey data) by the total area under aquaculture (ha) in Sundarbans. But this calculation creates the problem of double counting and gives a very high and incorrect number of workers (around 4.5 lakh workers, as against 7 lakh total agricultural workers). This could be due to including temporary farm workers along with the permanent workers in the total shrimp farm workers. These temporary shrimp farm workers, on an average, work only for 9 days in a year on a farm and then move on to another farm; therefore, they cannot be treated as workers (in accordance with the Census definition of workers) for a particular farm. However, these temporary workers could be counted as workers by the Census for a particular block on account of the total number of days they worked in a year.

2. This appendix is based on the Report submitted to IEG by the Centre for Environment and Development, Kolkata (2005). The household surveys in Canning and Minakhan blocks were conducted by the IEG team.

3. Especially those who own fishing boats often collect firewood from government forest without permit. The activity being illegal is either not admitted or denied.

4. Though most families own radio sets, the sets are not taken during fishing trips for fear of losing it in bad weather.
5. Firewood is usually sold in maunds; each maund equals 40 kilograms.

8

INTERNATIONAL DRIVERS AND LOCAL IMPACTS
RESPONSES AND INTERVENTIONS

RECAPITULATION

Earlier chapters have concluded that the expansion of processing for export and subsequent aquaculture-related economic activities has improved economic well-being of the stakeholders in the process. It has also had implications for land use, biodiversity, and water quality in the region. The combined effect has been a mixed one with positive and negative aspects. This trade-driven activity is expected to continue, with concomitant effects on different stakeholders. This chapter attempts to identify policy responses and interventions which augment the positive implications for different stakeholders. It shall also address the issue of identifying interventions which target the vulnerability of stakeholders—critical from welfare and income augmentation viewpoints and work towards sustainable use of the land and water resources of the region. Setting up ways and means for ensuring the applicability and feasibility of these responses by various agents at different scales of operations (local, state level, and national) would also be attempted in this synthesis chapter.

To recapitulate, we find from the analysis in earlier chapters that,

1) Incomes are indeed generated in the region due to shrimp farming activity. Human well-being indices based on a multi-dimensional approach to well-being provides an insight into changes brought about in the lives of different stakeholders. Trade-offs between aspects of well-being exist of which the

stakeholders are quite aware. To highlight the main aspects impacting different stakeholders:
- An export activity worth about Rs 5080 million at international prices results in income generation of Rs 6555 million in the year 2003–4. Every rupee of export results in income accrual of Rs 1.29. The distribution of this income generated is as follows: shrimp farmers (60.94 per cent), shrimp farm workers (13.77 per cent), PL collectors (8.01 per cent), transporters and retailers (13.81 per cent), and processing units with (3.71 per cent).
- Shrimp farmers in the areas of intensive expansion in land under aquaculture are better off than agriculturists in terms of per capita incomes. However, the security of these incomes is much lower in terms of frequency of normal harvests. Also, while agricultural farmers are not exposed to any conflict, shrimp farmers have a relatively lower level of social cohesion due to constant conflicts. Aquaculture farmers had to face higher conflict situations and more income insecurity than agricultural farmers, and
- PL collectors have increased incomes and reduced life insecurity, when compared to fishermen. They do not enjoy the level of income security, health security, and conflict-free livelihoods as salary and wage-earning households in the same area have. Alternative livelihoods based on the latter seem a viable option.

2) The scale at which the shrimp farms and processing units are operating presently is based primarily, on the lucrative export business. Any reduction in export opportunities or fall in profits will have a negative impact on the overall shrimp production activity. Once the export opportunities are exhausted or weakened due to any measure/reason, there will be a sharp reduction in the overall shrimp production.

3) All processing and production processes in aquaculture in the region are typically driven by the profit motive. During the nineties, aquaculture farming in West Bengal opted for semi-intensive production. It was only because of the outbreak of

disease due to intensive stocking that the vulnerability of incomes from such farming technology was revealed. A combination of economic rationale and legal action resulted in the adoption of an appropriate technology.

4) Full cost of using natural resources is not taken into account in the presence of short-run profit orientation and the simultaneous availability of cheap labour. This often leads to erosion of diversity. In this region in India, the huge expansion in land under aquaculture did not lead to the establishment of hatcheries to provide prawn seed, as would have been expected. This was due to the presence of adjacent water bodies populated with wild prawn seed together with the availability of cheap labour to collect these using crude methods of collection. Our study shows that these methods led to a decrease in biodiversity at several locations (as measured by an index based on time-series data). Further, results based on an econometric analysis of costs indicate that even if a biodiversity loss cost were to be assigned and seed prices to aquaculturists were to rise as a consequence, farmers would be able to absorb the rise, given the present structure of costs and the present price levels for their output. The elasticity coefficients also indicate that there exists a strong substitutability between land lease as inputs into aquaculture. In other words, a land-intensive aquaculture is indicated if biodiversity loss is to be averted.

5) Profit driven land-use change results in large tracts being converted to aquaculture. This conversion is far more pronounced than any other type of conversion (for instance from mangroves) recorded in the data analysis. The major factors driving land-use changes from mangrove and from agriculture to aquaculture are: encroachment of people (increasing population density), differentials in return on different kinds of land use including productivity of people in different types of activities. Over the period of thirty to forty years, a phased or sequential change in land use seems to be taking place. In the first phase, the conversion has been from mangroves to agriculture. In the second period, post-1986, which is the focus of our study, the conversion is from agriculture to aquaculture.

POLICY RESPONSES: REVIEWING THE CONCEPT

Responses refer to human actions, including policies, strategies, and interventions to address specific issues, needs or problems in various domains, that is, the society, the economy, and the environment. They are typically understood to stand for the act or 'state of doing'; responses can be individual or collective and may be designed to answer one or many needs at different scales of time and space.[1] When emanating from entities imbued with legal authority, responses become policies which guide economic activity; alternatively when decisions are taken by collective entities, they result in evolving social norms. Individual responses when expressed in markets collectively result in changed production structure.

In the context of aquaculture and ecosystem management in the Sundarbans, responses could be of several kinds. For example, they could be identified in terms of agents involved in the activities of shrimp seed collection, farming, transport, and processing (Central Government, West Bengal Government, District Administration of North and South 24 Parganas,) or in terms of disciplinary classifications such as legal, technical, institutional, economic, and behavioural. In other words, responses to any need or problem can be in the form of strategy, policy or directed measures. And the form they take depends on the particular social system in which the change is sought (Granovetter, 1985). The features of the social system will shape the nature, characteristics, and outcomes of responses. Institutional theory (Gladwin 1993) states that an institution is defined (by sociologists) as a collective and regulatory complex consisting of political and social agencies, which dominate other organizations through enforcement of laws, rules, and norms. Many, (for example, Oliver 1990), believe that institutional theory can accommodate interest-seeking behaviour when actor's responses to institutional pressures and expectations are not assumed to be invariably passive and conforming across all institutional conditions. Responses are conditioned by institutional set up and institutions themselves evolve to respond to change.

Any response, whether technical, legal or behavioural, has to recognize the context, cause, and criteria of its course. Such recognition will also enable the decision makers to examine and assess the effectiveness of responses in terms of attainment of the stated objectives. In general, the decision-makers or agents in society who take decisions to fulfil the desired objectives can choose different options in responding to a problem in ecosystem management. They can simply choose to ignore the problem they are facing if the intensity of problem is low or pressure to respond is not mounting or the decision makers themselves are insensitive to the problem which may arise due to ignorance, apathy or any other reason. They can effectively restrict or regulate the degradation by adopting and incorporating appropriate measures at each stage of the decision making process. Further, policies and interventions aimed outside the specific sector of aquaculture are also of interest as they may also lead us to well-targeted impacts.

In identifying specific policies in this chapter, we will focus our analysis on responses directly aimed at local environmental issues, as well as policies in other sectors such as economic, social and, legal instruments targeted at trade in aquaculture or at incentives for forest conservation. At times, the latter group (global trade and market structure-related responses) may have more impacts on the condition of ecosystems than the former. It is true that a large part of the fate of ecosystems is largely shaped by drivers of ecosystem change which may be global in nature. Integrated options or policies and measures that yield multiple benefits will also be highlighted. We will also attempt to evaluate the process of decision-making and the implementation of measures.

Often, negative trends in impacts on physical and economic characteristics of systems can be diagnosed, their causes identified, and possible cures designed at the local level more readily. However, the range of response options that can be meaningfully considered depends on the nature of the ecosystem problem, the linkages in the economic value chains and the social, economic, political, and institutional characteristics of the community in the area and the larger economy within which it functions.

This study constitutes one such example where changes in the Sundarbans are driven by trade, in particular by the increased culture and processing of shrimp for export. As discussed in Chapter 3, these changes were triggered by events at national and international levels.

The policy interventions relevant here therefore involve multiple actors, levels of organization, knowledge systems, and instruments of action and are complex. (Malayang, Thomas, and Kumar 2005).A wide spectrum of human responses can be envisioned involving different agents from individual, to community, national, and global levels.

We focus therefore on responses at various scales, ranging from local/village, national, to regional, and global scales. Special attention will be given to the interactions across scales, for instance, the impact international trade will have on the Sundarbans mangrove ecosystem management at national and local scales. Deforestation and loss of biodiversity are of global concern, but actions need to be taken at the local level. It is also at the lowest levels (household, village) that the societal domain and environmental domain interact. Here the pertinent issues to be examined shall be woven around the question: what are the linkages between the sets of responses at different levels? Does one link with, weaken or strengthen the influence of the other? Such an analysis will show pointers for the evolution of a more holistic policy towards aquaculture in the region.

EVOLUTION AND LISTING OF INDIVIDUAL AGENT FOCUSED AND INTEGRATED RESPONSES:[2] PRESENT AND FUTURE

This section analyses responses available to different agents. These emerge from an analysis of the inter-relation of trade in shrimp, the producers' response to it, and subsequent impact on ecosystems and rural poverty. Responses of different actors as they emerged in the decade of the nineties are discussed and possible directions for the future are indicated.

1. Exporting Units' level interventions—Units can internalize compliance costs imposed by international standards provided appropriate practices are followed and timely information is provided. Importing countries appropriately wish to ensure food safety through non-tariff measures of different kinds. These measures have become increasingly stringent over the 1990s as illustrated by the index set up in Chapter 3. An analysis of the processing units' cost structure suggests that, given the high international prices of shrimp, processing units can incur the required compliance costs. Such costs improve environmental conditions in and around processing units.

Further, in order to move towards eco-labelling and certification of different kinds, traceability of raw material to its source needs to be ensured. This will also ensure that banned chemicals and antibiotics which lead to subsequent cases of consignment cancellation/export rejections are not used in production of the raw material. Traceability also ensures that wrong practices at farm and market level which can lead to consignments being rejected are prevented.

Growing preference of international consumers for the prawn from this region implies that consumers are willing to pay for such measures. Simultaneously, the adaptive capacity of the exporting units to the international environment has exhibited itself and needs to be strengthened. The national government and industry associations have an important role to play in this. Fast transmission of accurate information on prevailing food safety standards and other requirements to exporters is essential to sustain incomes accruing in the region from this activity. The above policy intervention is of an enabling kind. Government can also ensure better hygiene and sanitary conditions in auction centers. Such measures shall also prevent the producers and workers lower down in the value chain from being exposed to unnecessary vulnerabilities caused by the volatile nature of the international market.

2. Shrimp farming targeted policy responses—choice of appropriate technology in shrimp farming has been ensured over time due to the right mix of legal measures and economic incentives. Diseases related with intensive stocking in aquaculture and subsequent income vulnerability reinforced the ban on intensive aquaculture. This is seen from the fact that local shrimp farmers have realized the potential threat to sustainable incomes from diseases and other ecological problems. They have devised the semi-intensive and improved traditional way of shrimp farming with lesser use of chemical fertilizers and pesticides. As Chapter 7 suggests, in the late nineties, farmers observed that intensive technology was subject to unpredictable disease and incomes could be highly fluctuating. Besides, farmers also faced a Supreme Court ban on this technology and settled for improved traditional technology. A combination of economic rationale and legal action resulted in the emergence of an appropriate technology. The continued adoption of such sustainable

aquaculture practices needs to be supported through appropriate provision of credit and market access.

However, continued concerns of aquaculture farmers are listed below:[3]

1. Lack of knowledge about scientific shrimp aquaculture,
2. Chamber to stock quality seeds,
3. Access to extension services for improved shrimp aquaculture,
4. Crisis of fresh water sources,
5. Water quality and disease outbreak,
6. Falling production and crop failure,
7. Lack of green cover,
8. By-catch and alternate employment for collectors,
9. Environmental degradation due to collection, and
10. Inadequate community and government initiative.

Multiple drivers of ecosystem change can be addressed through targeted and integrated responses, and earlier responses have sometimes not been well integrated.

3. In the region, the narrowly focused policy has clearly impacted selected relevant targets. Promotional intervention by fisheries department of the state government has accelerated the growth of export. The installation of the effluent treatment plant in the processing unit has reduced the pollution and in some cases pollution (water with excessive organic loads) from the ponds. Strict enforcement of the forest policy had definitely reduced the pace of mangrove deforestation as expected otherwise.

4. However, there has been a lack of integrated policies addressing the region as an economic and ecological entity as is evidently clear from the discussion in different chapters. For instance, an externality effect of the urban region of Kolkatta is on the quality of water released into the Sundarbans. This results in usage of bad quality water for washing of shrimps (and sometimes cooling) which many a time is the main cause of rejection of shrimp consignments. Further, quantity as well as quality of fish seeds in the region has drastically reduced due to discharge of urban untreated waste water. Discharge of this waste water affects the estuarine as well as marine ecosystem of the

Sundarbans. Deteriorating water quality has affected aquafarms both in terms of quantity as well as quality of produce. It is important to enforce effluent treatment and ensure better quality water release into the Sundarbans. Again, improved coordination between municipal authority, environment department, state pollution control board, and industry is needed to resolve the issue.

5. Continued lack of co-ordination across government departments and the failure to view the Sundarbans as a single ecological region and a world heritage site prevents meaningful policy response. Even when the Sundarbans Development Board was set up, its institutional structure (it was not headed by a 'senior' enough minister within the government) prevented its being effective to the required extent. As put forth by one of the stakeholders,[4] 'A Joint committee or a task force involving Central govt. and State govt. agencies along with market and exporters' associations is needed for implementation of existing or modified Code Of Conduct (COC). This may be instrumental in maintaining quality as well as proper price realization for the entire supply chain'.

6. The effectiveness of response options related to natural resources can be improved with coherence among different types of policies and the degree of collaboration among stakeholders.

As an instance, when the Forest Department is entrusted with protection of forests, the effect of economic factors on this mandate is not taken into account. Such neglect has expected results. An analysis of remote-sensing data for land use transformation in the region shows the disappearances of the forest although not to the same extent as claimed by some NGOs and other advocacy groups. Land-use change in the region for the period from 1986 to 2004 is explained in terms of differential rate of returns on paddy, aquaculture, and mangrove forest. Except Canning, all the blocks of the study region are showing depletion / degradation of mangroves. This is taking place inspite of the elaborate forest policy of the 1970s and the consolidated environmental policy of 1986 of the central government. The Forest department is seen as the sole protector of forests and people's aspirations for higher incomes are hardly taken into account. The effect of the aquaculture lobby or the big shrimp farmer and processing unit owner in the city is also ignored in the process. This lack of understanding of differential interest and priority of the stakeholders is the recipe for disaster as far as management and sustainable use of the mangrove of the Sundarbans is concerned.

A possible response is the introduction of some kind of land-zoning policy for aquaculture. As concluded by the WWF Workshop, 'Zonation for aquaculture is necessary to stop indiscriminate conversion of agricultural land. Inappropriate and unplanned establishment of shrimp farms has resulted in production failures, environmental degradation, land-use conflicts, and social injustice. Land zoning should factor in current impacts as well as future implications. Land zoning could not only help identify and demarcate safe zones but also the areas that should be under agriculture or other land uses'.[5]

7. Biodiversity Loss and Livelihood Generation—It is well-documented that subsequent to aquaculture expansion and in the absence of hatcheries, prawn seed collection is an important livelihood for large numbers of the poor. A well-developed transport network brings these PL collectors (post-larvae) to the farms. The crude collection methods result in loss of by-catch. Based on a trend analysis of biodiversity indices, this study in Chapter 5 concludes that a decline in biodiversity is emerging in some major collection centres.

The following policy options exist and may be considered:

(i) Internalize the cost of biodiversity loss in the aquaculture farming cost and set up a manner of spending amounts collected in a 'designated fund' on mangrove generation and more generally on biodiversity conservation. This is economically feasible as illustrated in Chapter 5. The challenge is to make it possible to implement administratively,.

(ii) Provide prawn seed through the 'hatcheries technology' and provide alternative livelihoods to prawn seed collectors. Such an option also requires extensive preparation by way of setting up of enabling social and legal frameworks, and

(iii) Legal bans such as that on the use of bag-nets which lead to by catch collection and loss may be enforced. Though the

Marine Fish Regulating Rules enforced it in 1998, the ban has been ineffective in the absence of alternate and viable means of livelihood.

In some cases, focused efforts of civil society and NGOs have brought the desirable change in the livelihood options for the people. Provision of livelihood options and non-conventional source of energy supply especially in areas like Chhotamullakhali has helped people in income generation with less impact on the stress of the ecosystems there. Development of small-scale tourism by involving local people (as in Bali Island) by Wildlife Protection Society of India (WPSI) and WWF has supplemented the effort of the state government. Co-operation between agencies such as the state, the private sector in identifying livelihood options is often facilitated by 'bridging organizations'.

Such efforts need to be extended on a much wider scale. In this context, the finding in Chapter 7 that salary-wage earner households in Chhotamullakhali have higher levels of well-being than prawn seed collectors is of great importance. Appropriate livelihood options need to be identified and adopted on larger scales in the short run. The chapter shows that these options have to be ones that ensure a secure daily income to the poor. In the longer run, augmentation of human skills and better connectivity with the mainland, together with development of new options such as through ecotourism provide feasible policy interventions.

8. In conclusion, when policy makers with different interests, experiences, and knowledge cooperate, the potential diversity of response options is enhanced. Society in the Sundarbans region, as in many other places, is driven by diverse interests and goals. Marginal people like prawn seed collectors try to maximize the daily catch of tiger prawn and in the process cause harm to biodiversity. Processing units want the continuous supply of the raw material (shrimp) for business where this supply constitutes more than 92 per cent of the total inputs. But the basic source of the supply is vulnerable if this demand by the processor cannot be met without adopting intensive farming, which while causing pollution and biodiversity loss, may be running the risk of white disease as well. Poor people in search

of prawn seeds damage other species resulting in biodiversity loss. People also try to penetrate the forest for temporary / permanent stay and land conversion causing the threat of forest clearing. Here diverse economic interests drive people. Faced with limited options, they are forced to make choices between these constituent ways in their best interest.

If these stakeholders receive facilitative help from the civil society and the government with inputs from diverse experiences and knowledge systems (especially of local people in dealing with protection of precious wildlife and forest etc), response strategies would prove far more effective in managing the ecosystem and its impact on the people's well-being. Such facilitative and enabling environments result in development, essentially an enhancement of peoples' ability to fulfil their aspirations in diverse ways.

CONCLUDING REMARKS: MOVING TOWARDS SUSTAINABLE DEVELOPMENT IN THE SUNDARBANS

The findings of our study (conducted between 2004 and 2006), reinforced by the findings from the outreach programme of the WWF completed in 2007 provide a picture of the kind of aquaculture which would result in sustainable development of the region. A mix of policy interventions in the form of legal measures and economic incentives is indicated. All these are targeted at ensuring that options available to different agents or stakeholders are enhanced in a manner which ensures their long-term sustainability and minimizes vulnerability. For aquacultural farmers, continued stress on appropriate technology and land tenure security supported by a land-zoning policy is indicated. Limiting the water pollution externality from urban human settlements and industries is another priority issue. For exporting units, is critical to be assured of quality control through traceability of raw material and timely information dissemination with regard to international standards. The more disadvantaged stakeholders such as the prawn seed collectors need to be enabled to pursue alternative livelihoods.

Understanding the links in the value chain for exporting aquacultural products and the economic, legal, and ecosystem linkages within which they function is critical towards the evolution of an inductive policy environment. This in turn requires a degree of understanding of regional imperatives which override the concerns of individual government and non-government agencies operating in the region. Each NGO and each government department has its own (even if well-intentioned and articulated) view. An overview of the region and its potential is somehow still missing. A regional authority such as the Sundarbans Development Board could have taken such an overarching view but, as discussed above, the opportunity seems to have been lost. Meanwhile, there exists an urgent need for a coming together of all stakeholders to deliberate on the unique nature of this region as an aquacultural hinterland and as a world heritage site endowed with a unique ecosystem. It is our earnest hope that the near future shall result in the emergence of processes and institutions which enable this and help in the movement towards sustainable development of the region.

NOTES

1. For a comprehensive treatment of the concept, see Millennium Ecosystem Assessment (2005), Policy Responses Group.
2. This section incorporates findings from Phase 2 of the project carried out by WWF in 2006–7. This phase involved a plan for outreach and the findings were presented at a stakeholder workshop held in Kolkatta on December 20, 2007.
3. Based on presentation at WWF December 07 workshop by N. Sarkar, representative, Progressive farmers.
4. The representative of the Market agents at the December 2007 Workshop.
5. Presentation by WWF at December 2007 Workshop.

Appendix-I

QUESTIONNAIRES

AQUACULTURE FARM QUESTIONNAIRE

Institute of Economic Growth
University of Delhi Enclave
North Campus
Delhi-110007

This information is required for the research study only. No part of this information will be used for any other purpose.

Serial number of the Questionnaire   Date of Interview

Name of the Interviewee
Designation of Interviewee
Name of the Interviewer
Location of the Farm

Remarks of the interviewer on the interviewee in terms of cooperation etc.

_____
_____
_____

Part I: General Details

1. Name of the owner of the farm:

2. Area of the production pond at the farm (bigha):

3. Depth of the pond (feet)

4. Is the Farm owned or leased by the farmer
   (1) _____ Owned
   (2) _____ Leased (number of years for lease)
   (3) _____ Government land
   (4) _____ Common land

5. Level of education of the farmer?
   (1) _____ Illiterate
   (2) _____ Literate (informal)
   (3) _____ Years of schooling if formal education

6. Since when has the farmer been into shrimp farming? (number of years at the site)

---

7. How much was the investment made by the farmer in setting up the farm?

		Investment (Rs)		
Year	Land	Construction of farm	Equipment	Total
Initial				
Present				

8. Do you incur any annual cost on pond maintenance? If yes, please specify the amount. _____ (Rs)

9. Was this farm an agricultural land/ paddy land or mangrove area prior to shrimp farming?
   (1) _____ Agricultural Land
   (2) _____ Paddy/ dhan land
   (3) _____ Mangrove area
   (4) _____ Common land
   (5) _____ Others (specify)

10. Why did you convert the land into a shrimp farm? (multiple answers can be given)
    (1) _____ Not fit for cultivation
    (2) _____ In lure of higher profits
    (3) _____ Agricultural crop failures
    (4) _____ Any other reason (please specify)

11. Name the months for shrimp cultivation in a year:
    (1) _____ (Peak season)
    (2) _____ (Lean season)

12. How many harvests in a year? _____ (numbers)

13. How long does one harvest take? _____ (number of months)

Part-II: Water Source and Quality

14. What is the source of water supply for the farm?
    (1) _____ Canal/river
    (2) _____ Estuaries
    (3) _____ any other source

15. What are the parameters that are tested in the water besides those mentioned below?(Look for other parameters in this visit). Multiple ticks expected.

pH level
Salinity
Metals (Zinc, Copper, Lead)
BOD
Any other

16. Do you treat water with chemicals before using it in grow out ponds?
    (1) _____ Yes
    (2) _____ No

If the answer to question no. 16 is yes, then ask question no.17

17. What are the chemicals/ antibiotics/ fertilizers that are used to treat the water used in shrimp ponds?

Chemicals/fertilizers/ medicines used	Total cost (Rs) in the last five years	Total cost (Rs) in the last ten years
Total		

18. How frequently water is exchanged with the common system (in how many days)? (water in-take from and discharged into creek/estuaries/river/ any other water body).

19. How many aquaculture farms are sharing intake water?
   _____numbers

20. How many aquaculture farms are discharging effluents into the common water body? _____numbers

21. Does the farm has a separate water supply/ drainage system?
    (1) _____ Yes
    (2) _____ No

22. How is the sludge disposed off from the farm and costs involved (Rs)?
    _____
    _____
    _____

Part III: Seed Quality and Source

23. What is the source of seed (shrimp PL)?

Source
1. Hatcheries of other states (specify)
2. Seeds collected from the wild(river/sea)
3. Local agent

24. What is the stocking density?(number of seeds in 000/bigha)
    (1) _____(Peak season)
    (2) _____(Lean season)

25. What is the seed cost in the following periods (Rs/000)
    (1) _____(Peak season)
    (2) _____(Lean season)

26. Is the seed collected from river legal? (does the government approve of this)?
    _____
    _____

27. How many days old PL is cultured in the farm? _____(days)

28. Has there been an increase in the price of shrimp PL in the last five years? If yes, please provide details.
   _____
   _____

29. Are there any tests/checks performed on shrimp juvenile/PL before they are stocked? If yes, then what are they?
   _____
   _____

Part IV: Feed Source and Quality

30. What is the source of feed given to shrimp PL?
    (1) _____ no formal feed
    (2) _____ agent from outside village/town
    (3) _____ local market
    (4) _____ any other source (please specify)

31. Feed details:

	Peak season	Lean season	Through out the year
Quantity of feed provided at one time *(kg/bigha)*			
Frequency of feeding per day (number of times)			
Feed requirement per harvest (in one cycle) *kg/bigha*			
Total Cost of the feed *(Rs/kg)*			

32. Has there been an increase in the price of feed in the last five years? If yes, please provide details.
    _____
    _____

33. 1 What is the composition of the feed given to shrimp PL?

Feed Inputs	Quantity *(kg/bigha)*	Price *(Rs/bag or Rs/quintals)*

33.2 What is the composition of the feed given to less than one month old shrimp PL?

Feed Inputs	Quantity *(kg/bigha)*	Price *(Rs/bag or Rs/quintals)*

34. Has the composition of the feed given to shrimp PL changed over the last five? years?
    (1) _____ Yes
    (2) _____ No

If answer to question no. 34 is yes, then ask question no. 35.

35. What is the composition of the feed given to shrimp PL now?

Feed Inputs	Quantity *(kg/bigha)*	Price *(Rs/bag or Rs/quintals)*

Part V: Production and Cost

36. What was the actual production in 2004? *(in kg/bigha or kg/harvest)*
    (1) _____ (Peak season)
    (2) _____ (Lean season)

37. What was the value of production in 2004? *(Rs)*
    (1) _____ (Peak season)
    (2) _____ (Lean season)

38. What has been the production trend over the last five years?

Year (1)	Production (Kg/ha) (2)	Reason for change (3)	Total loss (Rs) (4)	Total gain (Rs) (5)
1999				
2000				
2001				
2002				
2003				
2004				

Code: Column 3(Reason for change): Disease-1, change in stocking density-2, change in feed given to shrimp PL- 3, use of chemicals/fertilizer in the pond-4, change in the source of water supply-5

39. Where is the produce sold and to whom?
    (1) _____ local market
    (2) _____ Processing units
    (3) _____ self consumption
    (4) _____ any other (specify)

40. If the answer to question no.39 is local market then ask the following:

How far is the market *(kms)*			
What is the cost of transporting the produce to the market. *(Rs per trip)*	Peak season	Lean season	Total in one harvest
How many trips do you make to the market in the following seasons	Peak season	Lean season	In one harvest
Mode of transporting 　1. Boat 　2. Cart 　3. Van 　4. Other			
How much is the produce sold for *(Rs/kg)*	Peak season	Lean season	In one harvest

41. Has the selling price changed in the last five years? Please provide details.

Year	Selling price *(Rs/kg)*	Reason for change
1999		
2000		
2001		
2002		
2003		
2004		

42. Employment Details:

How many workers are employed on the farm (number)	Peak season	Lean season	Total
Wages paid *(Rs/person)*	Peak season	Lean season	Total
How many family members work on the farm (number)			
How many workers work permanently on the farm (through out the year)			
Wages paid *(Rs/person)*			

43. How much money do you save in case of good/normal harvest?

 _____ (Rs)

Part VI: Problems/Opportunities

44. What have been the major problems faced by the farmer at the site in the last five years?

 _____
 _____
 _____
 _____

45. The number of times that the farm has suffered from flooding problems in the last five years? _____ Numbers

46. Has the farm suffered from sedimentation problem in the past?
 (1) _____ Yes
 (2) _____ No

46.1. If yes, does it happen every year?
 (1) _____ Yes
 (2) _____ No

46.2. How do you overcome this problem and what are the costs involved. (For example, mud-digging/desiltation)

_____
_____
_____
_____

47.1 Number of times farm has suffered from disease problem in the last five years? _____ (number)

47.2 What has been the associated cost involved in overcoming this problem in the last five years (money spent on buying chemicals/ fertilizers/ medicines)? _____ (Rs)

48. Do you have an alternative source of livelihood in case of crop failure.? Please specify.

_____
_____
_____

Part VII: Subsidy and Financial Assistance

49. How many times has the farmer taken loan from the following sources?

	Amount taken (Rs) last year	Amount (Rs) taken in the last five years	Reason
1. Local money lender			
2. Government			
3. Relative/Family			
4. Post office/ Bank			

50. What is the estimated loan repayment cost (inclusive of the interest rate)? _____ (Rs)

51. Duration for which the loan is taken?
    (1) _____ less than 3 months
    (2) _____ 3-6 months
    (3) _____ 6-12 months
    (4) _____ More than 12 months

52. Has the government helped in providing inputs at subsidized rates?
    (1) _____ Yes
    (2) _____ No

If the answer to question no.52 is yes, then ask question no.53 and 54.

53. Government Assistance in providing the following at subsidized rates:

	Market Price *(Rs)*	Subsidized rate *(Rs)* provided by the government
Feed(*Rs/kg*)		
Seed(*Rs/000*)		
Fertilisers/chemicals/antibiotics *(Rs/kg)*		
Farm development		
Subsidy for purchase of water testing Kits/equipment for shrimp farms		
Setting up of hatcheries		

54. Costs involved in obtaining government's assistance.

Number of visits to government office to obtain assistance	
Money spent in each visit (transportation) *(Rs)*	
Time spent in a day in getting the assistance *(no. of hrs)*	
Money spent in obtaining subsidy *(Rs)*	
Labour days lost in obtaining the subsidy.	

55. Why could you not get government help/subsidy?

Too much paper work
Unable to establish contacts with the government or bank
Any other

Processing Unit Questionnaire

Institute of Economic Growth
University of Delhi Enclave
North Campus
Delhi-110007

This information is required for the research study only. No part of this information will be used for any other purpose.

Serial number of Questionnaire   Date of Interview
Name of the Interviewee
Designation of Interviewee
Name of the Interviewer

Remarks of the interviewer on the interviewee in terms of co-operation etc.:

Part I: General Details of the Unit

1.1. Name and address of the Establishment

1.2. Name of the Chief Executive (MD/Mg. Partner/Proprietor)

_____

_____

1.3. Is the processing plant owned or leased by the applicant?
   (1)_____Owned
   (2)_____Leased

1.4. If leased, name of the plant owner, plant name and address

_____

_____

1.5. When did the unit become operational?   _____ (Year)

1.6. What kind of activities does the unit undertake? *Mark only one answer.*
   (1)_____Pre-processing and processing
   (2)_____Processing only
   (3)_____Pre-processing, processing and export
   (4)_____Processing and export

1.7. What type of products do the unit process? *Mark only one answer.*
   (1)_____Shrimp only
   (2)_____Total marine products

1.8. Does the unit involve following operations? *Multiple answers can be given.*
   (1)_____Freezing and packing
   (2)_____Freezing only
   (3)_____Storage
   (4)_____Ice manufacture
   (5)_____All of the above

1.9. Could you please provide us the details of production of shrimp and total marine products for any year during each of the following time-periods?

Period	Quantity of Produce (tons)		Value of Produce (Rs Lacs)	
	Frozen Shrimp	Total Marine Products	Frozen Shrimp	Total Marine Products
1989–93				
1994–8				
1999–2003				

1.10. Could you please provide us the details of export of shrimp and total marine products for any year during each of the following time-periods?

Period	Quantity of Exports (tons)		Value of Exports (Rs Lacs)	
	Frozen Shrimp	Total Marine Products	Frozen Shrimp	Total Marine Products
1989–93				
1994–8				
1999–2003				

1.11. For how many months does the unit undertake processing in a year?
_____ Number of months

Part II. Economic Details of the Unit

2.1. What was the initial capital cost of the unit? _____ (Rs lacs)

2.2. What is the present capital cost of the unit? _____ (Rs lacs)

2.3. What type of raw materials do the unit use in processing of shrimp?

2.4. What percentage of shrimp and other marine products are bought from different sources for processing in the unit?

Shrimp (%)		Other marine products (%)	
Captured	Cultured	Captured	Cultured

2.5. What is the percentage share of captured shrimp and total marine products bought by the unit?

Supplier	Fishermen (%)		Agent (%)		Other Sources (%)	
Period	Shrimp	Total Marine Products	Shrimp	Total Marine Products	Shrimp	Total Marine Products
1989–93						
1994–8						
1999–2003						

2.6. What is the price (Rs/ton) of captured shrimp and total marine products bought by the unit?

Supplier	Fishermen		Agent		Other Sources	
Period	Shrimp	Total Marine Products	Shrimp	Total Marine Products	Shrimp	Total Marine Products
1989–93						
1994–8						
1999–2003						

2.7. What is the percentage share (%) of cultured shrimp and total marine products bought by the unit?

Supplier	Farmer (%)		Agent (%)		Other Sources (%)	
Period	Shrimp	Total Marine Products	Shrimp	Total Marine Products	Shrimp	Total Marine Products
1989–93						
1994–8						
1999–2003						

2.8. What is the price (Rs/ton) of cultured shrimp and total marine products bought by the unit?

Supplier	Fishermen		Agent		Other Sources	
Period	Shrimp	Total Marine Products	Shrimp	Total Marine Products	Shrimp	Total Marine Products
1989–93						
1994–8						
1999–2003						

2.9. Value of total inputs purchased for any year during each of the following periods:

Periods	Value (Rs Lacs)			
	Raw Material	Fuel	Water	Electricity
1989–93				
1994–8				
1999–2003				

Note: Raw Material: including shrimp and other raw materials, eg. preservatives etc, if any.

2.10. Information about personnel in the unit for any year during each of the following periods:

Periods	Employees & Workers (Numbers)			Wage Bill (Rs Lacs)			
	Skilled	Semi-skilled	Unskilled	Skilled	Semi-skilled	Unskilled	Total
1989–93							
1994–8							
1999–2003							

Part III: Environmental Regulations Details (National and International)

III. a. International Regulations:

3.1. To which country (ies) the unit is exporting shrimp and other marine products? *Multiple answers can be given.*
 (1) _____European Union (EU)
 (2) _____United States (US)
 (3) _____Japan
 (4) _____Any other

3.2. If the unit is exporting to EU, please tell us whether the unit is approved by Export Inspection Council (EIC) for exporting to EU?
 (1) _____Yes
 (2) _____No

3.2.1 If yes, when did the unit get EIC approval? _____ (Year)

3.3. If the unit is exporting to US and/or Japan, what kind of standards or regulations does the unit follow (for example, HACCP for US)? *Please specify.*

For US:
_____
_____

For Japan:
_____
_____

3.3.1 If yes, since when is the unit exporting to US and Japan?

_____ (Year)

3.4. Do you think that complying with EIC/HACCP or any other processing standards have led to an increase in total cost of processing (inclusive of official/paper work etc.)? If yes, what percentage of total cost you can assign to the compliance cost? Please explain.

_____

_____

_____

3.5. How many notifications from the importing country has the unit received in any year during each of the following periods for shrimp and other marine products?

Country	Number of Notifications relating to:				
	MRL	L&P	SPS	PPM	Total
EU:					
1989–93					
1994–8					
1999–2003					
US:					
1989–93					
1994–8					
1999–2003					
Japan:					
1989–93					
1994–8					
1999–2003					

Note: MRL – Maximum Residual Limits, L&P – Labelling & Packaging, SPS – Sanitary and Phyto-sanitary, PPM – Process and Product Method.

3.6. Details of compliance-related problems faced by the unit in exporting shrimp. Has the unit ever faced any kind of losses due to rejection/detention of shipments of shrimp by any importing country?

_____
_____
_____
_____

III.b. National Regulations:

3.7. How often the unit is visited by the WBPCB officials for monitoring pollution in a year? _____ (Numbers)

3.8. Does the WBPCB impose any kind of fine or penalty for non-compliance to water pollution standards caused by processing activity? Please give details:

_____
_____
_____

3.9. Did the unit ever receive any kind of subsidy, depreciation and tax concessions from the government for controlling pollution? If yes, please kindly provide details:

_____
_____

3.10. Who are the agencies or persons (either government e.g. EIC or private) with whom the unit has been interacting in connection with the pollution abatement or compliance with standards and the associated costs? Please provide details:

_____
_____
_____

3.11 Does the unit have to incur any legal expenses to deal with any kind of court cases of air and water pollution? Please describe the incidents and the total costs involved, if any.

_____
_____
_____
_____

3.12 Is rubbish (shrimp flesh, breading, soluble proteins, and carbohydrates) collected properly?
(1) _____ Yes
(2) _____ No

3.12.1 If yes, how is rubbish disposed off and what are the costs involved?

_____
_____
_____

3.13. Are the surroundings reasonably free from objectionable odours, smoke, dust, and other contamination?
(1) _____ Yes
(2) _____ No

3.14. Is drainage facility adequate?
(1) _____ Yes
(2) _____ No

3.15. If any other additive/chemical is used in processing, is it approved by the competent authority?
(1) _____ Yes
(2) _____ No

3.16. Is the water used for processing chlorinated to the accepted levels? (This is an EIC requirement).
(1) _____ Yes
(2) _____ No

3.17. Whether the unit tests finished products for heavy metals, antibiotics, pesticide residues, and biotoxins?
(1) _____ Yes
(2) _____ No

3.17.1 Does the HACCP Plan suitably address these requirements?
(1) _____ Yes
(2) _____ No

3.18. Is the own check system of the unit based on HACCP implemented?
(1) _____ Yes
(2) _____ No

3.19. Is the unit having in-house facilities for inspection and testing?
(1) _____ Yes
(2) _____ No

Part IV: Water Pollution Abatement in the Unit
(Questions 4.6 to 4.15.1 are for the ETP owning units)

4.1. Whether water used for processing in the unit meets the standards?
(1) _____ Yes
(2) _____ No

4.1.1 If yes, please tell us the source(s) of water for processing in the unit?
_____
_____
_____
_____

4.1.2 If no, how is the water made suitable for processing? Please provide information on the methods used and costs involved:
_____
_____
_____

4.2. What is the Water Cess as per the Water Cess Act currently paid by the unit:

Category	Industrial Cooling	Domestic	Biological Wastewater
Rate			
Amount per annum			

4.3. Average volume of wastewater generated per day and per annum by the unit:

Year	2003–4
Wastewater volume generated per day	
Wastewater volume generated per annum	

4.4. Is the unit having an efficient Effluent Treatment Plant (ETP)?
(1) _____ Yes
(2) _____ No

If no, please ask Questions 4.5 and 4.5.1 (for non-ETP units)

4.5. Where is the effluent released after treatment in ETP?
(1) _____ Municipal Drains
(2) _____ Any Municipal Treatment Plant
(3) _____ Open spaces
(4) _____ Any other place

4.5.1 What are the costs of disposing the effluent/sludge from ETP? Please give details:

_____

_____

4.6. When was the ETP installed? _____ (Year)

4.7. What was the value of Capital Stock of ETP in the year of its installation? _____ (Rs lacs)

4.8. What is the value of Capital Stock of ETP in the present year?
_____ (Rs Lacs)

4.9. Number of People employed in ETP: (2003-04)

Workers & Employees	Numbers	Wage Rate (Rs/hr) or (Rs/month)	Annual Wage Bill (Rs)
Engineers/technologists			
Skilled			
Unskilled			
Total			

4.10. What is the annual expenditure of the ETP in the year 2003-04?

Costs	Quantity or unit	Value *(Rs lacs)*
Maintenance		
Material		
Energy/fuel		

4.11. What is the Wastewater Handling Capacity of the ETP?
_____ (Volume)

4.12. What type of water treatment does the unit undertake?
    (1) _____ Primary
    (2) _____ Secondary
    (3) _____ Primary and Secondary
    (4) _____ Primary, Secondary, and Tertiary

4.13. Average Characteristics of Untreated wastewater (Influent Quality)

	2002–3	2003–4
BOD		
COD		
TSS		
PH		
TDS		
Oil and Grease		

4.14. Average Characteristics of Treated wastewater (Effluent Quality)

	2002–3	2003–4
BOD		
COD		
TSS		
PH		
TDS		
Oil and Grease		

4.15. Where is the effluent released after treatment in ETP?
    (1) _____ Municipal Drains
    (2) _____ Any Municipal Treatment Plant
    (3) _____ Open spaces
    (4) _____ Any other place

4.15.1. What are the costs of disposing the effluent/sludge from ETP? Please give details:

_____
_____
_____
_____
_____

HOUSEHOLD QUESTIONNAIRE

Institute of Economic Growth
University of Delhi Enclave
North Campus
Delhi-110007

This information is required for the research study only. No part of this information will be used for any other purpose.

Serial number of Questionnaire	Date of Interview
Name of the Interviewee	
Designation of Interviewee	
Name of the Interviewer	

Remarks of the interviewer on the interviewee in terms of co-operation etc:

_____
_____
_____

1. House Description and Characteristics:

Ownership (Code)	Type of structure (Code)
Religion of Head	Caste of head (where relevant Code)
Village	Community Development Block
District	

Code: Ownership: rented-1, owned-2. Type of structure: pucca-1, kutcha-2, semi-pucca-3.
Religion: Hindu-1, Muslim-2, Christian-3, Other-4. Caste: General (upper caste)-1, SC/ST/OBC (lower caste)-2, other-3.

2. Household Roaster:

ID No.	Name (Start with respondent)	Relationship to head (Code)	Sex (M-1, F-2)	*Education (number of years of formal education)	Member(s) currently studying outside village-1, in village-2, vocational course-3	Age (Years)	Currently married Yes-1, No-2	Whether earning member Yes-1, No-2
(1)	(2)	(3)	(4)	(5)	(6)	(7)	(8)	(9)
1								
2								
3								
4								
5								
6								
7								
8								
9								
10								
11								
12								

Code: Column 3(Relationship to head): self-1, spouse of head-2, child/spouse of child-3, grandchild-4, other relation-5.
Note: * Illiterate-0, Literate-99 (in case of non-formal education)

3. How many household members are contributing to the household income in cash or kind?
   In Cash: _____ (Number)
   In kind (by collection for self-consumption) _____ (Number)

4. Livelihood of Earning Members

ID No.	Name of Earning Member (Including self-employed on farm)	Sex (M-1, F-2)	Occupation/ Trade	Number of days employed in a month (for all categories)	Where employed (in island of residence: 1/outside: 2)	Distance commuted everyday for work	Mode of transport
(1)	(2)	(3)	(4)	(5)	(6)	(7)	(8)
1							
2							
3							
4							
5							
6							

Codes: Column 4:
Cultivation on own farm-1, Agricultural labourer-2, Shrimp Farming-3, Shrimp PL collection-4, Shrimp farm labourer-5, Fishing-6, Hired by fisherman-7, Trade in village-8, Trader to nearby town-9, Employment in nearby town-10, Other-11. Identify others in pretesting

Column 5: Enter approximate days for farmers, fishermen, traders, workers etc.

Column 8: Boat-1, on foot-2, bicycle-3 bus-4 (two of the above could be entered)

(A) For Agricultural Farmer Households

5. Landholding and cultivation: Unit-acres or a local unit that can be converted into acres

S. No.		Unit	Quantity
1.	Total agricultural land owned		
2.	Net sown area in last season		

5.1. What percentage of the land you cultivate is irrigated, if any?
_____(%)

5.2. Was any of your land not cultivable in the last season due to flooding, If yes how much? _____ (ha)

6. Information on crops cultivated during the past 12 months:

1. Name of crop	2. Which crops did you sell, if any? (Tick if sold)	3. Quantity sold *(Kg)*	4. Price at which sold *(Rs)*	5. Which crops did you buy, if any (Tick if bought)	6. Quantity bought *(Kg)*	7. Price at which bought *Rs Per kg*
1. Rice (Dhan)						
2.						
3.						
4. Other						

7. Do you own or operate a shrimp farm
   (1) _____ Yes
   (2) _____ No

(B) For Shrimp Farmer Households
Question nos. 8 to 12.3 for those who say yes to question no. 7.

8. Details regarding land/pond under shrimp farming
Area of the pond: (in acres or local units): _____

Owned/ leased in: _____

(Code: Owned-1, Leased in-2, Short period lease-2a, Long period lease-2b (Long period is more than 10 years), On Common land without leasing agreement-3, Do not know-4.

9. How many harvests of shrimp (on average) do you get in a year?

_____(Number)

10. Questions on shrimp farm: Inputs

1. Name of inputs	2. Source of input	3. Quantity purchased annually (if any) (Kg)	4. What was the value of purchase (Rs)	5. From whom purchased
Seed				
Feed				
Insecticide				
Water				
Any other				

Code for Column 2: Purchased-1, Collected-2
Code for Column 5: Local people-1, Local Agent-2, Agent from outside village/town-3, Any other-4

11. Questions on shrimp farm: Output Disposal per harvest

1. Disposal of outputs	2. Quantity sold per harvest (in kg.)	3. Price per kg. (For different sale categories) (Rs Per kg.)
Self-consumption		
Sale in village market		
Sale in W.B. market		
Sale to processors		
Sale to export processors		

12. How many years have you operated/owned the shrimp farm:

_____ (Number)

12.1. How many years has harvest been normal/above normal:
_____ (Number)

12.2. Have there been any disputes with respect to the farm/pond in the last five years?
(1) _____ Yes
(2) _____ No

12.3. If yes to 12.2 then, how were they resolved:

Resolved By	Tick (√) against right option	Time taken in resolution (Hrs./days)	Number of meetings
(1) Local panchayat			
(2) Party representative			
(3) District or other court			
(4) Mutual consent			

(C) For Fishermen
Question Nos. 13 to 17.4 for households with fishing as livelihood

13. Where do you go for fishing?
(1) _____ River
(2) _____ Creeks

13.1. How many days in a month do you go fishing?   _____ (days)

13.2. How many months in a year do you go fishing?   _____ (months)

14. Equipment Used for fishing

1. Name of equipment	2. Date of purchase	3. Price when purchased (Rs)	4. Remaining Life (in years)	5. How purchased?	6. If Hired, cost of hiring per day/per month
Boat					
Fishing Net					
Any other					

Note: Column numbers 2 to 5 are for owned equipment, 6 for hired equipment.
Code for column 5: Self-funded-1, Borrowing from money lender-2, Borrowing from sales agent/trader-3, Borrowing from family-4, Any other-5.

15. Questions on fishing: Output Disposal per harvest

1. Disposal of outputs	2. Quantity sold per harvest *(in kg.)*	3. Price per kg. (For different sale categories) *(Rs Per kg.)*
Self-consumption		
Sale in village market		
Sale in W.B. market		
Sale to processors		
Sale to export processors		

16. How often in the last five years have you encountered heavy storms while out fishing? _____ Number per year

1.1. How many accidents at fishing place have you heard of in the last five years? _____ Number

1.2. Is there adequate information on weather forecast at fishing place?
   (1) _____ Yes
   (2) _____ No

16.3. Any health related problems from water exposure?
_____
_____
_____
_____
_____

17. Do you encounter any problems in selling your output?
   (1) _____ Yes
   (2) _____ No

18. Is there any sales co-operative in your area?
    (1) _____ Yes
    (2) _____ No

18.1. Are you a member of such a co-operative?
    (1) _____ Yes
    (2) _____ No

19. Any recent conflict related to fishing rights in your area?
    (1) _____ Yes
    (2) _____ No

19.1. If there were any conflicts please give their details.

_____
_____
_____
_____

(D) For Shrimp PL Collectors

20. Where do you go for shrimp PL collection?
    (1) _____ River
    (2) _____ Creeks

20.1. How many days in a month do you go for shrimp PL collection?
    _____ (days)

20.2. How many months in a year do you go for shrimp PL collection?
    _____ (months)

21. Equipment Used for shrimp PL collection

1. Name of equipment	2. Date of purchase	3. Price when purchased (Rs)	4. Remaining Life (in years)	5. How purchased?	6. If Hired, cost of hiring per day/per month
Boat					
Fishing Net					
Any other					

Note: Column numbers 2 to 5 are for owned equipment, 6 for hired equipment. Code for column 5: Self-funded-1, Borrowing from money lender-2, Borrowing from sales agent/trader-3, Borrowing from family-4, Any other-5.

22. Questions on shrimp PL Collection: Output Disposal per harvest

1. Disposal of outputs	2. Quantity sold per harvest *(in kg.)*	3. Price per kg. (For different sale categories) *(Rs Per kg.)*
Self-consumption		
Sale in village market		
Sale in W.B. market		
Sale to processors		
Sale to export processors		

23. How often in the last five years have you encountered heavy storms while out for shrimp PL collection?   _____ Number per year

23.1. How many accidents at shrimp PL collection place have you heard of in the last five years?   _____ Number

23.2. Is there adequate information on weather forecast at shrimp PL collection place?
   (1) _____ Yes
   (2) _____ No

23.3. Any health-related problems from water exposure?

_____

_____

_____

_____

_____

24. Do you encounter any problems in selling your output?
    (1) _____ Yes
    (2) _____ No

25. Is there any sales co-operative in your area?
    (1) _____ Yes
    (2) _____ No

25.1. Are you a member of such a co-operative?
    (1) _____ Yes
    (2) _____ No

26. Any recent conflict related to fishing rights in your area?
    (1) _____ Yes
    (2) _____ No

26.1. If there were any conflicts, please give their details.

_____

_____

_____

_____

_____

(E) For Wage/salary earning households

27. Are you working for a wage/ salary?
    (1) _____ Yes
    (2) _____ No

27.1. If yes to question 27, then fill the following table:

1. Nature of Employment	2. Nature of job*	3. Daily/Monthly salary (Rs)	4. Distance of place of work from home (in kms.)	5. Probability of getting work**	6. Uncertainty of payment***	7. Risks at Work****
On agricultural farm						
On shrimp farm						
With fisherman						
In government office in town						
As building / construction worker						
As teacher						
Other						

*Code for Column 2: Permanent-1, Contractual-2, Temporary-3.
** Code for Column 5: Rank probability on a scale of 0.1 to 1 with 1 being complete certainty
***Code for Column 6: Uncertainty of payment: Rank from 1 to 5 with 1 being complete certainty.
**** Code for Column 7: Risk of life in storm-1, Health risk mild-2, Health risk severe-3, Risk of being stranded in town-4, Risk of punishment by forest guards-5.

27.2. Would you change your employment if you could, even if you got the same income?
(1) _____ Yes
(2) _____ No

(F) Question Numbers 28 onwards for all households

28. Do any member(s) of your household collect any items
    (a) from common land : Yes / No
    (b) from common water resources: Yes/ No

28.1. If yes, name five important items collected (Enter names of five items here)
    (a) Common Land Resources: _____
    (b) Common Water resources: _____

29. Details of the collection from land/ forests in particular months, if there are any?

Name of product collected	Months for collection with number of days per month	Reason for collection (Code)	Place of collection (Code)	Average time spent in collection (Hrs./Day)
Fodder/grasses				
Fuel wood				
Honey				
Medicinal/ aromatic herbs				
Others (specify)				

Note: The five items mentioned in Q21 should be included here.
Code: Reason for Collection: regular self-consumption for household-1; self-consumption/sale only during a particular period of distress-2; regular sale activity to traders-3; collection for own (member) processing at co-operative 5; Any other
Place of collection: village forest-1, govt. forest-2, private land (farmland)-3, common land-4.

30. Details of the collection from water (river, creek) in particular months, if there are any?

Name of product collected	Months for collection with number of days per month	Reason for collection (Code)	Place of collection (Code)	Average time spent in collection (Hrs./Day)
Prawn Seed				
Fish				
Medicinal/aromatic herbs				
Others (specify)				

Note: The five items mentioned in Q 21 should be included here.
Code: Reason for Collection: regular self-consumption for household-1; self-consumption/sale only during a particular period of distress-2; regular sale activity to traders-3; collection for own (member) processing at co-operative-5; Any other
Place of collection: village creek-1, river-2, private ponds-3, common water ponds-4.

31. If you sell, whom do you sell the products to from among the following?
    (1) _____ A co-operative
    (2) _____ Traders
    (3) _____ Other households in the village

32. If traders, where do they come from?
    (1) _____ Stay in village
    (2) _____ Come from outside the village

33. Does your household get some credit facility from these traders?
    (1) _____ Yes
    (2) _____ No

34. How far is the nearest market/selling place for the produce (for direct sales to traders)? _____ (Km)

35. Any problems that you encounter in collecting products from land and water?

Nature of Problem	Working Time lost per month in problem (Days per month)	Cost incurred in solving problem (Rs Per month)	No. of incidents in last year (Number)	No. of visits per incident (to government office, panchayat, doctor)
Harassment by Forest Department				
Risk to life from sharks				
Health / Skin problem from standing in water				
Conflict with other groups				
Any other				

36. Price received per unit sales to:

Produce	Price from (Rs/Kg)		
	Co-operative	Traders	Other households
Prawn Seed			
Fish			
Fodder/grasses			
Honey			
Fuelwood			
Bamboo			
Leaves			
Medicinal/aromatic herbs			
Others (specify)			

37. Would you give up collecting from the water and land if you were given the same amount as income every month?
    (1)_____ Yes
    (2)_____ No

37.1. Would you give up collection for sale but still collect for self-consumption?
    (1)_____ Yes
    (2)_____ No

38. Has any member of your household changed his/ her source of livelihood in the last five years
    (1)_____ Yes
    (2)_____ No

38.1. Has anyone in your family added a new source of earning to his/her livelihood in the last five years?
    (1)_____ Yes
    (2)_____ No

Question 39 for households in which either of questions 38 and 38.1 elicit a positive reply.

39. Livelihood Options

1. Present Occupation
2. Earlier Occupation
3. Reason for change*

*Code for Row 3: More financial security: 1 More life security: 2 Less risk to health:3 Less friction with group conflict (including with forest department): 4
Consequent on higher level of education: 5 Less commuting distance: 6 Higher wage: 7
If more than one reason, rank them and mention by highest rank (beginning with 1 as highest rank)

(G) General Household Status

40. Monthly household expenditure on food and non-food items (*all figures in Rs*)

Item ➔	Food	Education of children	Education of adult	Travel for work	Travel for education	Health	Recreation and entertainment	Other (any other major head)
Total Expenditure (Rs)								

41. Where does your drinking water principally come from? *Mark only one answer.*
    (1) _____ Piped water
    (2) _____ Tubewell/handpump
    (3) _____ Tank/pond
    (4) _____ River/canal/lake/spring
    (5) _____ Other

42. What is the main source of lighting in your house? *Mark only one answer.*
    (1) _____ Electricity
    (2) _____ Gobar gas/oil/kerosene
    (3) _____ Other (specify)

43. What kind of fuel do you use for cooking? *In case of multiple answers, rank according to importance.*
    (1) _____ LPG
    (2) _____ Electricity
    (3) _____ Kerosene
    (4) _____ Coal
    (5) _____ Firewood
    (6) _____ Other (specify)

44. Does your household own any of the following items? Yes-1, No-2.

Item	Bicycle	Motorcycle/scooter	Motor car	Refrigerator	T.V.	Telephone

45. Does any household member read a newspaper or listen to the news on radio/TV daily?
    (1) _____ Yes
    (2) _____ No

46. Household borrowings during the last 12 months:

Source of borrowing	Whether borrowed money from Yes—1, No—2	Purpose of borrowing (Code)	Total amount borrowed (Rs)
Bank			
Post office			
Private lender			
Family			
Other			

Code: Column 2 (purpose of borrowing): agricultural farm business-1, shrimp farm business-2, fishing equipment purchase: 3 purchase of residential land or building-4, marriage-5, medical-6, education-7, debt repayment-8, Other-9.

47. Has there been any major law and order problem in your area in the past five years?
   (1) _____ Yes
   (2) _____ No

47.1. If yes, what is the cause for it: (tick two and give ranking):
   (1) _____ Fishing rights
   (2) _____ Forest entry/ rights
   (3) _____ Drinking Water
   (4) _____ Caste/ Community/ Religion
   (5) _____ Market Sale and Traders
   (6) _____ Any Other

REFERENCES

Agarwal, B., Jane Humphries, and Ingrid Robeyns (eds). 2007, *Capabilities, Freedom and Equality: Amartya Sen's work from a Gender Perspective.* Oxford University Press (OUP), New Delhi, India.

Alkire, S. (2007), 'Choosing Dimensions: The Capability Approach and Multidimensional Poverty', CPRC Working Paper 88, OPHI, Oxford.

────── (2002), 'Dimensions of Human Development', *World Development,* Vol. 30(2), pp. 181–205.

────── (2001), *Valuing Freedoms: Sen's Capability Approach and Poverty Reduction*, Oxford University Press, Oxford.

Anand, S., and A. Sen (2000), 'The Income Component of the Human Development Index', *Journal of Human Development,* pp. 83–106.

Appffel-Marglin, F., and S. Marglin (1990), *Dominating Knowledge: Development, Culture, and Resistance,* Clarendon Press, Oxford.

Berkes, F., J. Colding, and C. Folke (eds) (2003), *Navigating Social-Ecological Systems: Building resilience for Complexity and Change*, Cambridge University Press, Cambridge, UK.

Berndt, E.R. and D.O. Wood (1975), 'Technology, Prices, and the Derived Demand for Energy', *The Review of Economics and Statistics*, Vol. LVII (3), pp. 259–68.

Bhagwati, J., and V.K. Ramaswamy (1963), 'Domestic Distortions, Tariff and the Theory of Optimum Subsidy', *Journal of Political Economy,* Vol. 71, pp. 44–50.

Blasco, F. (1977), *Outlines of Ecology, Botany and Forestry of the Mangals of the Indian Subcontinent Wet Coastal Ecosystem,* Scientific Publishing Company, Oxford Elsevier.

Bilsborrow, R. and Geores, M. (1995), 'Population, Land Use and the Environment in Developing Countries: What Can we Learn From Cross National Data?', Ch. 8, in Katrina Brown and David W. Pearce (eds), *The Causes of Tropical Deforestation: The Economic and Statistical Analysis of Factors Giving Rise to the Tropical Forests,* University of British Columbia Press, Vancouver.

Bureau of Applied Economics and Statistics (2002), *District Statistical Handbook*, Kolkata.

Centre for Environment and Development (2005), *Household Survey—Chhoto Mollakhali, Sundarbans*, A commissioned report for the project 'Trade, Environment and Rural Poverty', Institute of Economic Growth, Delhi.

Chambers, R. (1997), 'Responsible Well-being—A Personal Agenda for Development', *World Development*, Vol. 25(11), pp. 1743–54.

Chompitz, K.M. and D.A. Gray (1996), 'Roads, Land Use and Deforestation', *World Bank Economic Review*, Vol. 10(3), pp. 487–512.

Chopra, K. and G.K. Kadekodi (1999), *Operationalising Sustainable Development: Economic Ecological Modelling for Developing Countries*, Indo-Dutch Studies on Development Alternatives, Sage Publications, New Delhi, London.

Chopra, K. and R.N. Agarwal (1999), *Environment Regulations as Trade Barriers for Development Countries: the Case of Some Indian Agriculture Exports*, Discussion Paper Series No. 7/99, Institute of Economic Growth, p. 22.

Chopra, K. (1985), 'Substitution and Complementarity between Inputs in Paddy Cultivation', *Journal of Quantitative Economics*, Vol. 1(2), pp. 315–32.

Costanza, R., L. Waignerm, C. Folke and K.G. Maler (1993), 'Modelling Complex Ecological Economic Systems: Towards an Evolutionary Dynamic Understanding of People and Nature', *Bio Science*, Vol. 43, pp. 545–55.

Cropper, M., C. Griffiths, and M. Mani (1999), 'Roads, Population Pressures, and Deforestation in Thailand, 1976–89', *Land Economics*, Vol. 75, pp. 58–73.

Daly, G. (1997), *Nature's Services*, Island Press, Washington D.C.

Danda, A.A. (2007), *Surviving in the Sundarbans: Threats and Responses, An Analytical Description of Life in an Indian Riparian Commons*, Doctoral Thesis, University of Twente, The Netherlands.

Dasgupta, P. (2001), *Environment and Human Well-being*, Oxford University Press, Oxford.

Dasgupta, P. and K.G. Maler (eds.) (1997), *The Environment and Emerging Developmental Issues*, Vols. 1 and 2, Clarendon, Press Oxford.

——— (1994), *Poverty Institutions and the Environmental Resource Base*, World Bank Environment Paper No. 9, The World Bank, Washington, D.C.

Directorate of Census Operations (2001), *Census of India (2001), Provisional Population Totals*, Series–20, Paper 3.

Directorate of Census Operations (2001), *Census of India, District Census Handbooks and Village and Town Directory, Primary Census Abstract*, North 24 Paraganas and South 24 Paraganas. Series–20, Part XII– A.

———— (1991), *Census of India, District Census Handbook and Village and Town Directory, Primary Census Abstract*, North 24 Paraganas and South 24 Paraganas Districts, Series–26, Part XII– A.

———— (1991), *Census of India–1991, Provisional Population Totals*, Series–26, Paper – 3.

Duraiappah, A.K. and A. Israngkura (2000), *Farm Permits and Optimal Shrimp Management in Thailand: An Integrated Inter-temporal and Spatial Planning Model* CREED Working Paper No. 35, IIED, London and IES, Amsterdam.

Duraippah, A.K. (1996), *Poverty and Environmental Degradation: A Literature Review and Analysis*, CREED Working Paper Series No. 8, London, IIED.

FAO (2003), *Aquaculture: Not Just an Export Industry*, Food and Agriculture Organization of the United Nations, Rome, Italy.

———— (1996), *Agriculture Production Statistics*, 1984–94, FAO Fisheries Circular 815 (Rev8), Food and Agriculture Organization of the United Nations, Rome, Italy.

Gasper, Des (2003), *Nussbaum's Capabilities Approach in Perspective: Purposes, Methods and Sources for an Ethics of Human Development*, Working paper series No. 379ISS, The Hague.

———— (1996), Needs and Basic Needs: *A Clarification of Meanings, Levels and Different Streams of Work*, Institute of Social Studies Working Paper No. 210, The Hague.

Gladwin, T.N. (1993), 'The Meaning of Greening: A Plea for Organizational Theory', Kurt Fischer and Johan Schot (eds), *Environmental Strategies for Industry*, Island Press, Washington, D.C.

Global Biodiversity Assessment (1995), Cambridge University Press, Cambridge.

Goldar, B., S. Misra, and B. Mukherji (2001), 'Water Pollution Abatement Cost Function: Methodological Issues and an Application to Small-scale Factories in an Industrial Estate in India', *Environment and Development Economics*, Vol. 6, pp. 103–22.

Government of India. *Economic Survey (2004–05)*, Economic Division, Ministry of Finance.

———— (2002), 'Status of Shrimp Farming in West Bengal', *Aquaculture Authority*, Vol. 1(1), Ministry of Agriculture.

Government of India (2001), *Report of the Commission for Agricultural Costs and Prices*, Department of Agriculture and Co-operation, Ministry of Agriculture.
——— (2001b), *Staff Papers on Poverty*, Ministry of Statistics and Programme Implementation.
———, *Economic Survery (various issues)*' Economic Division, Ministry of Finance.
Government of West Bengal (2004), *Administrative Report (1999–2000 to 2003–04)*, Sundarban Development Board (SDB), Sundarban Affairs Department, Kolkata.
——— (2004), *West Bengal Human Development Report (WBHDR)*, Development and Planning Department, Government of West Bengal.
Government of West Bengal, *Economic Review 2001–02*, State Planning Board, Kolkata.
Granovetter, M. (1985), 'Economic Action and Social Structure: The Problem of Embeddedness', *American Journal of Sociology*, Vol. 91(3), pp. 481–510.
Gunawardena, M. and J.S. Rowan (2005), 'Economic Valuation of a Mangrove Ecosystem Threatened by Shrimp Aquaculture in Sri Lanka', *Environmental Management*, Vol. 36(4), pp. 535–50.
Hamilton, K. and J.M. Hartwick (2005), 'Investing Exhaustible Resource Rents and the Path of Consumption', *Canadian Journal of Economics*, Vol. 38(2), pp. 615–21.
Hamilton, K. and M. Clemens (1999), 'Genuine Savings Rate in Developing Countries', *World Bank Economic Review*, Vol. 13(2), pp. 333–56.
Henson, S., Ann-Marie Brouder, and Winnie Mitullah (2000), 'Food Safety Requirements and Food Exports from Developing Countries. The Case of Fish Exports from Kenya to the European Union', *American Journal of Agricultural Economics*, Vol. 82(5), pp. 1159–69.
Humphrey, D.B. and J.R. Moroney (1975), 'Substitution among Capital, Labor, and Natural Resource Products in American Manufacturing', *Journal of Political Economy*, Vol. 83(1), pp. 57–82.
Kadekodi, G.K. and A. Mishra (2003), *Trade Related Environmental Regulations: Some Lessons from India*, Paper read at Seminar at Indira Gandhi Institute of Development Research, July, Mumbai.
Kates, R.W., B.L. Turner, and W.C. Clark (1990), 'The Great Transformation', pp. 1–17, in B.L. Turner, W.C. Clark, R.W. Kates, J.F. Richards, J.T. Matthews, and W.B. Meyer (eds), *The Earth is Transformed by Human Actions*, Cambridge University Press, England.
Kumar, A. and P. Kumar (2003), 'Food Safety Measures: Implications for

Fisheries Sector in India', *Indian Journal of Agricultural Economics*, Vol. 58(3), July–September.

Kumar, P. (2005), Assessment of Economic Drivers of Land Use Change in the Urban Ecosystems of National Capital Region (NCR), Delhi, mimeo, Institute of Economic Growth, Delhi.

Malayang, B., H. Thomas and P. Kumar (2005), *Responses to Ecosystems Change and Their impact on Human Well Being*, Sub Global Assessment, MA, Island Press, Washington D.C.

Marine Products Export Development Authority (various issues) *The Marine Products Export Review*, Ministry of Commerce. Government of India.

Markusen, J.R. (1999), 'Location Choice, Environmental Quality and Public Policy', in van den Bergh (ed.), *Handbook of Environmental and Resource Economics*, Edward Elgar Publications, Cheltenham.

McGillivray M. (2005), 'Measuring Non-economic Well Being Achievement', *Review of Income and Wealth*, p. 51.

Mehta, R. and J. George (2005), *Food Safety Regulation Concerns and Trade— The Developing Country Perspective*, Macmillan India Ltd, Delhi.

Millennium Ecosystem Assessment [MEA] (2005), *Ecosystems and Human Well-being: Policy Responses*, Findings of the Responses Working Group, Island Press, Washington D.C.

────── (2003), *Ecosystems and Human Well-being—A Framework for Assessment*, Island Press, Washington D.C.

Ministry of Environment and Forests (1996), *Impact of Mass Collection of Prawn Seeds in Mangrove Ecosystem of Sunderbans Biosphere Reserve*, S.D. Marine Biological Research Institute, West Bengal, India.

Mitra, A. (2005), *Study of the Evaluation of Fin Fish Juvenile Loss due to Wild Harvest of Tiger Prawn Seeds from Coastal West Bengal*, A commissioned report by the Department of Marine Science, University of Calcutta for the project 'Trade, Environment and Rural Poverty', Institute of Economic Growth, Delhi.

Mitra, R. and S. Hazra (2005), 'Agricultural Vulnerability at Bhitarkanika Wildlife Sanctuary', Orissa, School of Oceanographic Studies, Jadavpur University, Kolkata, Paper presented at the INSEE conference, June 2005 in IGIDR, Mumbai.

Morgenstern, R.D., W.A. Pizer, and J.S. Shih (2001), 'The Cost of Environmental Protection', *The Review of Economics and Statistics*, Vol. 83(4), pp. 732–38.

Narayan, D., R. Patel, K. Schafft, A. Rademacher, and S. Koch-Schullem (2000), 'Voices of the Poor: Can Anyone Hear Us', Oxford University Press for the World Bank, New York.

Narayan, D., R. Chamber, M.K. Shah, and P. Petesch (2000b), 'Voices of the Poor: Crying Out for Change', Oxford University Press for the World Bank, New York.

Nelson, G.C., V. Harris, and S.W. Stone (2001), 'Deforestation, Land Use, and Property Rights: Empirical Evidence from Darien, Panama', *Land Economics*, Vol. 77(2), pp. 187–205.

Nelson, G.C. and D. Hellerstein (1997), 'Do Roads Cause Deforestation in Southern Cameroon', *Applied Geography*, Vol. 17, pp. 143–62.

North, D.C. (1990), *Institutions, Institutional Change, and Economic Performance*, Cambridge University Press, Cambridge, UK.

Nussbaum, M.C. (2001), *Upheavals of Thought—The Intelligence of Emotions*, Cambridge University Press.

―――― (2000), *Women and Human Development: The Capabilities Approach*, Cambridge University Press, Cambridge.

Nussbaum, M.C. and Amartya Sen (eds) (1993), *The Quality of Life*, Clarendon Press, Oxford.

Odum, E.P. (1985), 'Trend Expected in Stressed Ecosystems', *Bio Science*, Vol. 35, pp. 419–22.

Oliver, C. (1990), 'Determinants of Inter Organisational Relationships: Integration and Future Directions', *Academy of Management Review*, Vol. 15, pp. 241–65.

Osberg, L. (2003), 'An Index of Economic Well Being for Selected OECD Countries', *The Review of Income and Wealth*, pp. 291–316.

Pfaff, A.S.P. (1999), 'What Drives Deforestation in the Brazilian Amazon? Evidence from Satellite and Socioeconomic Data', *Journal of Environmental Economics and Management*, Vol. 37, pp. 26–43.

Primavera, J.H. (2000), *Integrated Mangrove—Aquaculture Systems in Asia*, Aquaculture Department, Southeast Asian Fisheries Development Center, Tigbauan, Philippines.

―――― (1993), 'A Critical Review of Shrimp Pond Culture', *Reviews in Fisheries Sciences*, Vol. 1, pp. 151–201.

Pritchard, L., J. Colding, F. Birkes, U. Svedin, and C. Folke (1998), *The Problem of Fit Between Ecosystems and Institutions*, International Human Dimensions Programme on Global Environmental Change.

Qizilbash, M. (1996), 'Capabilities, Well-being and Human Development: A Survey', *Journal of Development Studies*, p. 33.

Rees, W. and M. Wackernagel (1994), 'Ecological Footprint and Appropriated Carrying Capacity', in A.M. Jansson et al. (eds), *Investing in Natural Capital*, Island Press, Washington D.C., pp. 362–90.

Rosamond L.N., Rebecca J. Golburg, H. Mooney, M. Beveridge, J. Clay,

C. Folke, N. Kautsky, J. Lubchenco, J. Primavera, and M. Williams (1998), 'Nature Subsidies to Shrimp and Salmon Farming', *Science,* Vol. 282, pp. 883–84.

Sen, A.K. (1999), *Development as Freedom,* Knopf Press, New York.

Seto, K., C.E. Woodcock, C. Song, X. Huang, J. Lu, and R.K. Kaufmann (2002), 'Monitoring Land-use Change in the Pearl River Delta Using Landsat TM', *International Journal of Remote Sensing,* Vol. 23(10), pp. 1985–2004.

Shyam, S.S., C. Sekhar, K. Uma, and S.R. Rajesh (2004), 'Export Performance of Indian Fisheries in the Context of Globalisation', *Indian Journal of Agricultural Economics,* Vol. 59(3), pp. 448–64.

Stewart, F. (1996), 'Basic Needs, Capabilities, and Human Development', in A. Offer (ed.), *Pursuit of the Quality of Life,* Oxford University Press, Oxford.

Subba Rao, D.V. (1994), *Sustainable Development: An Integrated Optimal Land-use Approach,* Institute of Economic Growth, Working Paper No. E/162/94, Delhi.

Sugden, R. (1993), 'Welfare, Resources and Capabilities: A Review of Inequality Re-examined by Amartya Sen', *Journal of Economic Literature,* Vol. 31, pp. 1947–62.

Tholkappian, S. (2005), 'Environmental Regulation: Hidden Costs and Empirical Evidence', *Economic and Political Weekly,* February 26, pp. 856–59.

Ulph, A.M. (1999), *Trade and Environment: Selected Essays of A.M Ulph,* Edward Elgar Publications, Cheltenham.

United Nations Development Programme (UNDP) (2004), *Human Development Report,* 2004. *(http//hdr.undp.org/reports/global/2004)*

UNEP–IISD (2004), *Human Well-being, Poverty and Ecosystem Services,* Nairobi.

Wackernagel, M. and Rees, W. (1996), *Our Ecological Footprint: Reducing Human Impacts on the Earth,* New Society, Gabriola Press, British Columbia, Canada.

World Bank (2006a), *Aquaculture: Changing the Face of Waters, Meeting the Promises and Challenge of Sustainable Aquaculture,* Agriculture and Rural Development, Washington D.C.

——— (2006b), *Where is the Wealth of Nations? Measuring Capital for the 21st century,* The World Bank, Washington D.C.

——— (2001a), *Globalization, Growth and Poverty: Building an Inclusive World Economy,* Oxford University Press, Washington and New York.

——— (2001b), *World Development Report,* Washington and New York.

World Resources Institute (Undated), *Farming Fish: The Aquaculture Boom*. Fact Sheet WRI, Washington D.C.
WWF (2007), *Policy Dialogue for Promoting Sustainable Shrimp Aquaculture in the Sundarbans*, December 2007, Kolkata.
UN Comtrade database.

www.fda.gov
www.jetro.go.jp
www.mhlw.go.jp
www.mohfw.nic.in
www.foodmarketexchange.com
www.indiastat.com
www.westbengalstat.com
www.fao.org

NAME INDEX

Agarwal, B., 195n1
Agarwal, N. 60n1
Alikre, S. 171, 172
Anand, S. 171
Apffel–Marglin, F. 165n6

Berkes, F. 19n5
Bhagwati, J. 60n1
Bilsborrow, R. 134
Birkes, F. *see also* Folke et al.
Bogstrom 10

Chamber, R. *see also* Narayan D. et al
Chaudhary, Jayshree Roy 20n13
Chompitz, K.M. 135
Chopra, K. 60n1
Clark, W.C. *see also* Kates R.W. et al.
Colding, J. 19n5
Colding, J. *see also* Folke et al.
Costanza, R. 10
Cropper, M. 137

Daly, G. 174
Dampier, William 88
Dasgupta, Partha 2, 171, 173, 174, 195n3
Duraiappah, A.K. 173

Folke, C. et al. 8, 9, 11, 19n5, *see also* Costanza, R. et al.

Gasper, Des 19n6, 171
Geores, M. 134
George, J. 15, 60n1
Goldar, B. 126n6
Gray, D.A. 135
Griffiths, C. 137
Gunawardena, M. 9

Hahn 209
Harris, V. 138
Hein, L. 196
Hellerstein, D. 138
Hodges, Lt. 88
Humphries, Jane 195n1
Humprey 126n5

Kadekodi, G.K. 60n1
Kates, R.W. et al. 132
Kumar, Praduman 15, 61, 139–40, 209
Kumar, Anjani 15, 61

Malayang, B. 209
Maler, K.G. 2, *see also* Costanza, R. et al
Mani, M. 137

Marglin, S. 165n6
Markusen, J.R. 60n1
McGillivray, M. 171
Mehta, R. 15, 60n1
Mishra, A. 60n1
Misra, S. 126n6
Mitra, Abhijit 126n1, 126n7, 195n10
Morgenstern, R.D. 115
Moroney, J.R. 126n5
Mukerji, B. 126n6

Narayan, D. et al. 5, 174
Nelson, G.C. 138
Nussbaum, M.C. 19n6, 171, 195n1

Odum, E.P. 10
Osberg, L. 171

Petesch P. *see also* Narayan D. et al.
Pfaff, A.S.P. 136
Pizer, W.A. 115
Pritchard, L. *see also* Folke et al.

Qizilbash, M. 171

Rajesh, S.R. *see also* Shyam, S.S. et al.
Rees, W. 10
Robeyns, Ingrid 195n1
Robeyrs 195n1
Rowan, J.S. 9

Sekhar, C. *see also* Shyam, S.S. et al.
Sen, Amartya 4, 171
Shah, M.K. *see also* Narayan D. et al.
Shih, J.S. 115
Shyam, S.S. et al. 15
Stewart, F. 19n6
Stone, S.W. 138
Sugden, R. 19n6, 195n2
Svedin, U. *see also* Folke et al.

Turner, B.L. *see also* Kates R.W. et al.

Ulph, A.M. 60n1
Uma, K. *see also* Shyam, S.S. et al.

Wackernagel, M. 10

Subject Index

antibiotics 97, 99–100, banning of 210
'Approach to Capability' 172
aquacultural products, chain for exporting 215–16
aquaculture, 6, 8; ban on intensive 210; bans on use of bag-nets 213; and damage to juvenile finfish community 108; and ecological cost of 9–11; environmental impacts of 7–9; FAO definition of 12; farmers 211; industry and trade regulations 14–16; lacunas of modern 9; landzoning policy for 213; in Sundarbans 108; and world economy 6–7
aquatic production, in Asian countries 13; in India 13; profit motive in 205–6
aquatic resources destruction by discard of by-catch 109–10
Aratdars 99
Asian financial crisis 22

Bayesian Maximum Likelihood 144
Bengal Basin 88
biodiversity 108, see also deforestation; erosion, cost of 119; loss, cost of 113–17, 213; loss and innovative policies 125–6; loss and livelihood generation 313–14;
Brackish Water Fish Farmer's Development Agency (BFFDA) 24
Breusch Pagan Lagrangian Multiplier (LM) 55
Bureau of Indian Standards (BIS) 63, 103

'capabilities' approach 4; to poverty 171
Carp production 13, 16
'carrying capacity' 10
chemicals 8, 117; banning of 210, see also Pharmacologically Active Substances
Chhotamullakhali 214
China as producer of shrimp 27–30, 35
Chotto Mollakhali (Gosaba Block) Survey 198–202
civil society 214, 215, see also NGOs
coastal, communities 8; water pollution 7
Code of Conduct (COC) 212
Codex Alimentarius Commission (CAC) 63

Competitor's Import Price (CPR) 54
Critical Control Point (CCP) 14
crude oil price increase 22

deforestation and loss of biodiversity 209
Department of Fisheries, Aquaculture, Aquatic Resources, and Fishing Harvest (DFAARFH) 24–7
Development, definition of 4
Diamond Harbour 111–12
Digha Tourist centre 111

eco-labelling 62, and certification 210
ecological crop loss, indices of 110–13
ecological crop 109
Econometric methods 144
economic and social institutions 3
economic liberalization 21
ecosystems 2–3; change, drivers of 211; estuarine 109; kinds of 19n1; as 'natural capital' 1; services 170–6; of Sundarbans 5
ecotourism 214, *see also* ecosystems services
error component model (ECM) 50
EU regulations 62
Exchange Rate Movements 21–3
export activity 205
Export Inspection Agency (EIA) 64, 103
Export Inspection Council (EIC) 64, 103
export units, level interventions 209; and processing 94–5

export, of marine products from India 79; of shrimp 85, *see also* aquaculture; products 14; of shrimp from West Bengal 45–9

fish products 14; global trade in 7
Fisheries, as source of employment 91
flood hazard dummy 136–7
Food and Agriculture Organization of United Nations (FAO) 63
Food safety standards (FSS) 14–15
food-borne hazards 14
foreign direct investment (FDI) 140
foreign exchange earnings 85
Forest department and forest protection 212
frozen shrimp exports 39–43; and EU 37, 48–9; from India 76–9; Japan as destination of 37, 48; to USA 37, 48; from west Bengal 36–8

'ghost acreage' 10
good life, poor's idea of 5
Good Manufacturing Practice (GMP) 14

habitat fragmentation 7
Harinbhanga-Taimanga-Ichamati 88
'hatcheries technology' 213
Hausman test 55
Hazard Analysis Critical Control Point (HACCP) systems 14–15
Hoogly-Matla estuary 88
Human Development Index (HDI) 173
human well-being 4–6, 171–3;

Subject Index 271

determinants of 174;
  indices 181–4; MA definition
  of 5
'human' capital 2
icthyoplankton 108
India and world shrimp economy
  27
Indian, import price INDP 54, 56;
  Food Safety Regulations 63
Indices of Dominance, of
  Diamond Harbour 130; of
  Junput 131; of Sagar South
  130
Indices of Evenness, of Diamond
  Harbour 128; of Junput 129;
  of Sagar South 129
individual agent focus and
  integrated responses 209–15
institutions 2
international food safety standards
  and Indian seafood industry 61
investors, private 101

Japanese Regulations 63
Junput 111–12, *see also* Indices
  of Dominance; Indices of
  Evenness

Labelling and Packaging (L&P)
  64, 70, 71, 103
Lagrangian Multiplier (LM) test
  56
land-use change 132–40, 212; bio-
  physical drivers 134; drivers
  of 139; estimation of 155; in
  Indian Sundarbans 140–7;
  kinds of 147–59; profit driven
  206; property right on 138; in
  Sundarbans 159;
Liberalized Exchange Rate
  Management System (LERMS)
  22
Local Markets (Domestic
  Consumption) 102
local transporters 102

mangrove 8–9, 19–20n11,
  86–7; anthropogenic pressures
  on ecosystems of 132;
  aquaculture-driven conversion
  of 9; forests 17–18; as reserved
  88; survival of 111
Marine Fish Regulating Rules
  Enforcement, 1998 214
Marine Products Export
  Development Authority
  (MPEDA) 60n2, 95, 103
marine products exports 23–7,
  30–6; and shrimp exports from
  India 30–6
Maximum Residual Limit (MRL)
  64, 69, 71, 103
meen [or post-larvae (PL)] collector
  85, 136, 205; and fishermen in
  Gosaba 188–90; salary-wage
  earners in Gosaba 190–2;
Mexican crisis 22
Millennium Ecosystem Assessment
  (MA) 5, 174
Ministry of Health and Family
  Welfare (MOHFW) 63, 64

National Codex Contact Point
  (NCCP) for India 63, 64
natural capital 2
natural resources use, cost of 206
New Economic Policy 21, 23
NGOs 212, 214, 216
Non-tariff Measures (NTM) 14,
  49–51, 52, 53

Non-Tariff Measures, Index of 67–8; and Real Value of Exports 80–4

off-farm post-harvest production links 101–2

Pharmacologically Active Substances 65–6, *see also* antibiotics
policies 133, 207–12; development project 137; and export of the shrimp 24; for mitigating Biodiversity loss 125; and myopic policy choice 9; on trade 3
population increase 136, 156
Post-Harvest Production Links 101
poverty 173, 170–6; definition of 4
PPM: Product processing methods 69
prawn, international consumers for 210; seed collectors 98–9
Processing Units 179, 180–1
Product and Process Method (PPM) 103

Random Effects Model 55
'real wealth' and 'real savings' 2
resource consumption and waste discharge 10
rupee depreciation 22
Russian crisis 22

Sagar South 111–12
Sanitary and Phyto-sanitary Measures (SPS) 14–15, 61–2; rating of 69–70, 71–2

Shannon Weaver species diversity index 112
shrimp aquaculture 27; trade in 12–14
shrimp farming 24, 108; commercial 27, 85; and policy responses 210–11
shrimp production 170; in India 30; input suppliers 99–100; profile 176; Stakeholders in 93–4
shrimp, cultivation 8; culture activity 85; culture and resource use 192–5; export 176–7; farm owner-cum-worker 96–8; farm workers 179; farmer 95–6, 186–8, 205, 213; in Canning and Minakhan 184–6; PL Collectors 179; and salmon 12
Shrimp-farming technology, Supreme Court ban on 192, 210
Species Diversity Indices of Diamond Harbour 127; Junput 128; Sagar South 127;
Sundarbans 11, 16–18; aquaculture and as mangrove tiger-land 89; dampier-hodges line 88; ecosystem management 16, 207; Indian 8, 176–81, 196–7; mangrove tiger-land of globe 88; migrated population in 18; political system in 100–1; region 86–90; socio-economic profile of 90–3; and sustainable development 215–16
Sundarbans Development Board 212, 216

Subject Index

Sundarbans National Park as UNESCO world heritage site 16

Technical Barrier to Trade (TBT) Agreement 14–15, 61–2
tiger prawn seeds, demand for 108–11; sampling stations 111
Transport and Trade Margins 180

United Nations Development Program (UNDP) 173
urbanization 7
Uruguay Round GATT Agreement 23
US Food and Drug Administration (FDA) 62
US Regulations 62–3, 72

water quality, deterioration of 212
Water-Feed Cost 118–19
well-being 170–6
wild fish, damage to 11
Wildlife Protection Society of India (WPSI) 214
World Health Organization (WHO) 63
World Trade Organisation (WTO), agreement on Technical Barriers to Trade (TBT) 103; India as founder member of 23; Indian marine products and agreements of 15; technical Barriers to Trade Enquiry Point 103
Worldwide Fund for Nature (WWF) 213, 214, 215